AN APPROACH TO 1G MODELLING
IN GEOTECHNICAL ENGINEERING
WITH SOILTRON

ADVANCES IN GEOTECHNICAL ENGINEERING AND
TUNNELLING

11

General editor:

D. Kolymbas

University of Innsbruck, Institute of Geotechnics and Tunnel Engineering

In the same series (A.A.BALKEMA):

1. D. Kolymbas (2000), *Introduction to hypoplasticity*, 104 pages, ISBN 90 5809 306 9

2. W. Fellin (2000), *Rütteldruckverdichtung als plastodynamisches Problem, (Deep vibration compaction as a plastodynamic problem)*, 344 pages, ISBN 90 5809 315 8

3. D. Kolymbas & W. Fellin (2000), *Compaction of soils, granulates and pow-ders - International workshop on compaction of soils, granulates, powders*, Innsbruck, 28-29 February 2000, 344 pages, ISBN 90 5809 318 2

In the same series (LOGOS):

4. C. Bliem (2001), *3D Finite Element Berechnungen im Tunnelbau, (3D finite element calculations in tunnelling)*, 220 pages, ISBN 3-89722-750-9

5. D. Kolymbas (General editor), *Tunnelling Mechanics, Eurosummer-school, Innsbruck, 2001*, 403 pages, ISBN 3-89722-873-4

6. M. Fiedler (2001), *Nichtlineare Berechnung von Plattenfundamenten (Nonlinear Analysis of Mat Foundations)*, 163 pages, ISBN 3-8325-0031-6

7. W. Fellin (2003), *Geotechnik - Lernen mit Beispielen*, 230 pages, ISBN 3-8325- 0147-9

8. D. Kolymbas, ed. (2003), *Rational Tunnelling, Summerschool, Innsbruck 2003*, 428 pages, ISBN 3-8325-0350-1

9. D. Kolymbas, ed. (2004), *Fractals in Geotechnical Engineering, Exploratory Workshop, Innsbruck, 2003*, 174 pages, ISBN 3-8325-0583-0

10. P. Tanseng (2005), *Implementations of Hypoplasticity and Simulations of Geotechnical Problems*, in print.

11. *This book.*

12. L. Prinz von Baden (2005), *Alpine Bauweisen und Gefahrenmanagement (Alpine Construction Methods and Risk Management)*, 243 pages, ISBN 3-8325-0935-6

13. D. Kolymbas & A. Laudahn, eds.(2005), *Rational Tunnelling, 2nd Summerschool, Innsbruck, 2005*, 304 pages, ISBN 3-8325-1012-5

An Approach to 1g Modelling in Geotechnical Engineering with Soiltron

Andreas Laudahn

University of Innsbruck, Institute of Geotechnics and Tunnelling

E-mail: andreas.laudahn@uibk.ac.at
Homepage: http://geotechnik.uibk.ac.at/

The first three volumes have been published by Balkema
and can be ordered from:

A.A. Balkema Publishers
P.O.Box 1675
NL-3000 BR Rotterdam
e-mail: orders@swets.nl
website: www.balkema.nl

Titelbild:
Model test of a foundation with Soiltron;
left: model container, right: test evaluation (solid line)
and prototype measurements (dashed line)

Bibliographic information published by Die Deutsche Bibliothek

Die Deutsche Bibliothek lists this publication in the Deutsche National-
bibliografie; detailed bibliographic data is available in the Internet at
http://dnb.ddb.de.

ISBN 3-8325-1072-9

ISSN 1566-6182

Logos Verlag Berlin
Comeniushof, Gubener Str. 47,
10243 Berlin
Tel.: +49 030 42 85 10 90
Fax: +49 030 42 85 10 92
INTERNET: http://www.logos-verlag.de

Contents

Preface

Complex problems in geotechnical engineering and soil mechanics can be investigated with physical modelling, where the structures are reproduced in a geometrically n-times smaller scale and the occuring forces and deformations are measured. Problematic is the reduction of pressures and the subsequent change of the mechanical behaviour of the model soil.

In this thesis the similarity between barotropy and pyknotropy was used to model prototype soil behaviour. Soiltron is a prototype soil, which is treated with light and soft additives to achieve the same relative density in the model at an n-times smaller pressure as in the prototype. The stiffness and strength of the new material can be influenced to simulate prototype material behaviour by the control of the soil density. Thus, the use of that soil as material in small scale $1g$ models is possible.

At the Institute of Geotechnical and Tunnel Engineering of the University of Innsbruck a model soil material was *not* available in sufficient quantity. Therefore, a large quantity of quartz sand was purchased and its mechanical behaviour was investigated in detail with triaxial test devices using local strain gauges. The specimen deformations were inspected with evaluation of remote measurements using PIV. Systematic and random errors have been avoided or have been taken into account.

Appropriate regressions have been found to relate pressure and density to the properties of the soil and to calculate the required density of Soiltron. The appropriateness of Soiltron is verified in two demonstration model tests.

This work can be seen as foot step to strike a new path in physical modelling.

I am deeply grateful to my advisor Prof. Dimitrios Kolymbas for his invaluable guidance, courtesy and constructive suggestions throughout my study. I also deeply appreciate the advice and encouragement by Prof. Gerd Gudehus.

Special thanks go to my colleagues at the institute for their support and for the very pleasant working atmosphere. My sincere thanks also my colleagues K. Sosna and J. Boháč in Prague for their generous assistance. Gratefully I acknowledge also the financial support of the Austrian-Czech Research Grant 'Aktion'.

Finally, my love to my family: Claudia and Johanna. Thank you for tolerating my frequent absence from home.

Chapter 1

Physical models in geotechnical engineering

Large scale field tests are cost-intensive, time consuming and, due to the variability and complexity of soil conditions and construction processes, difficult to arrange and interpret. Therefore, alternatives like laboratory model tests (physical modelling) or numerical modelling are often chosen. In geotechnical engineering physical modelling is needed not only for design but also for research purposes, such as the validation of numerical models.

Soil behaviour is controlled by pressure, which is often given by gravity. The importance of gravity induced stresses in the behaviour of soil can be understood from Fig. 1.1, where the stress distribution in the soil, a) in the prototype as well as b) and c) in n times smaller models are shown. In the $1 : n$ centrifuge model, Fig. 1.1b, the vertical stress is reproduced with the same magnitude if the gravitation is increased n times, n being the geometrical scaling factor. In the $1g$ small-scale model Fig. 1.1c, the gravity induced stresses are n times smaller.

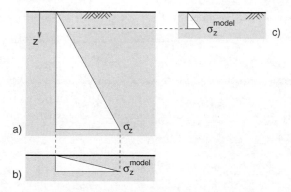

Fig. 1.1: Vertical stress distribution due to gravity: a) prototype, b) small scale $(1 : n)$ centrifuge model, c) $1g$ small scale $(1 : n)$ model

The influence on mechanical behaviour by the stress level is the so-called *barotropy*, and the influence of density is the so-called and *pyknotropy*.[1] These soil features limit the capability of small scale models under normal gravitation to capture realistically the behaviour of soils, see section 1.1 and 1.5.1. In this study, a possibility to overcome this difficulty is proposed.

1.1   Important features of soils

1.1.1   Barotropy and pyknotropy of soils

Barotropy is the dependence of mechanical behaviour on pressure-level. *Pyknotropy* is the dependence of mechanical behaviour on density.

Barotropy is shown in Fig. 1.2, where stress ratio and volumetric strains of triaxial tests of a soil with *equal initial density* at low and high confining pressures are plotted. The graph shows idealized plots of the evolution of the stress ratio σ_1'/σ_3' in a triaxial test. For low confining pressures the curve exhibits a peak value, marked with P in the sketch, and decreases beyond the peak with increasing strain (softening). The volume decreases (contractancy) in the beginning and increases with loading (dilatancy).

The behaviour of the same soil under high confining pressure is markedly different: the friction angle is lower and there is no pronounced peak in the stress-strain curve, the volumetric behaviour is contractant. Strength and stiffness increase nonlinearly with confining pressure.[2] Note that the peak friction angle is not a soil constant.

Idealized pyknotropy is shown in Fig. 1.3, where the stress ratio and volumetric strains of a dense and a loose soil in the triaxial test *at the same confining pressure* are plotted versus the axial strain. A dense soil shows a higher strength and stiffness than a loose one. The stiffness of a dense soil is decreasing after reaching a peak value (softening). The behaviour is dilatant after initial contractancy. A loose soil does not exhibit a pronounced peak in the normalized stress-strain curve and shows after an initial contractancy no volume increase.

---

[1] Furthermore, the mechanical behaviour of soil is nonlinear and irreversibile, cf. section 1.1.2, what has minor influence on the geometrical scaling in this study.

[2] Kolymbas, D.: Geotechnik – Bodenmechanik und Grundbau, Springer-Verlag 1998, p. 155 et sqq.

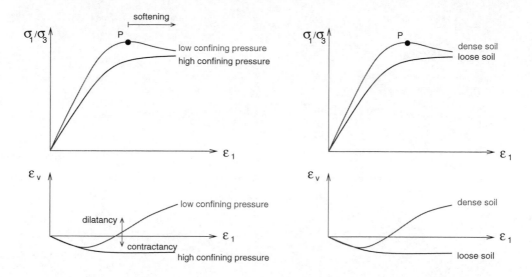

Fig. 1.2: Sketch to the barotropy of a soil

Fig. 1.3: Sketch to the pyknotropy of a soil

Fig. 1.4 shows experimental results of triaxial tests on Ottendorf-Okrilla-Sand of nearly the same initial density at five different confining pressures. The tests were conducted by the author and students[3,4] in the soil mechanical laboratory of the Charles University in Prague using sophisticated devices, such as local displacement transducers and hydraulic pressure/volume controllers.[5]

The shear strength (maximal deviatoric stress) increases nonlinearly with confining pressure, as can be seen in Fig. 1.6 the shear strengths form a curved envelope.

To reproduce the pressure dependency of soil behaviour realistically in a downscaled model (1:n model), the model can be either placed into a centrifuge, see section 1.4, and the same stress level is obtained by increasing g:

$$g = \omega^2 r \; ; \; \gamma = \varrho g \; ; \; \sigma = \gamma h \; .$$

Or one can take advantage of the similarity in soil mechanical behaviour (strength, stiffness and dilatancy) of loose soils under low stresses and dense soils under high stresses, as proposed here with the use of Soiltron.

---

[3]Sosna, K.: Pevostní charakteristiky směsí písku (Stiffness characteristics of sand mixtures), MSc. thesis at the Charles University Prague, 2003

[4]Hiegelsperger, M.: Nichtstandardmäßige Triaxialversuche (Non-standard triaxial tests), MSc. thesis at the University of Innsbruck, 2004

[5]A detailed description of the devices and the evaluation of the tests is given in sections 3.1-4.1

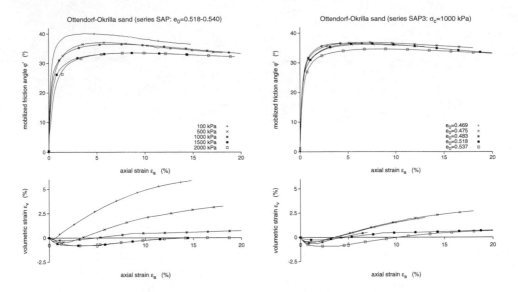

Fig. 1.4: Barotropy: triaxial tests with different confining pressures and nearly the same initial density, tests SAP1D, SAP2C, SAP3D, SAP4C and SAP5B, described in detail in section 4.1

Fig. 1.5: Pyknotropy: triaxial tests with the same confining pressure and varying initial density, tests SAP3I, SAP3H, SAP3G, SAP3D, SAP3B (e_0 from 0.469 to 0.537), described in detail in section 4.1

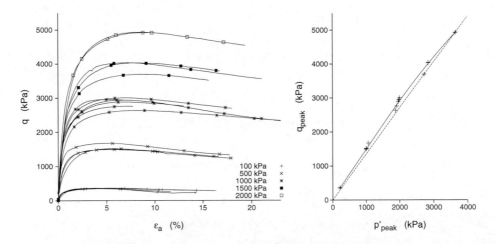

Fig. 1.6: Stress dependence of mechanical behaviour of soils; left: stress deviator q of triaxial tests on Ottendorf-Okrilla-sand, see section 4.1; right: shear strength vs. mean stress, peak values of these tests (dotted line: linear increase of shear strength with mean stress; continuous line: average of measured values, marked with +)

The aim of this study is to utilise the effects of barotropy and pyknotropy on the mechanical behaviour of soils for small scale laboratory modelling of geotechnical problems without using centrifuges. In prototype structures, high pressures dominate in the soil compared to a small scale model in laboratory, cf. Figures 1.1a and 1.1c. As can be seen from the comparison of Figures 1.2 and 1.3 the mechanical behaviour of a loose soil is similar to a denser soil under higher stress.

> *Hence, it should be possible to simulate the stress-strain and volumetric behaviour of a prototype soil with a looser soil in a geometrical reduced model without scaling of gravity.*

Further important features of soil behaviour are nonlinearity, irreversibility, hardening, softening and yielding. This will be outlined in the following.

1.1.2 Nonlinearity and irreversibility

In many engineering problems the stress distribution in investigated bodies is predicted from the assumption of elasticity, than stresses can be superimposed. But for soils the relationship between stresses and strains (stiffness) is incrementally nonlinear. The nonlinearity and the related irreversibility has to be considered if complex loading paths are investigated in physical or numerical models.

Irreversibility of strains is shown in Fig. 1.7 e.g., where the result of a triaxial test with Ottendorf-Okrilla sand (test SACP1) with quasi-static unloading and reloading cycles is plotted. Triaxial test SACP1 was conducted strain-controlled, i.e. the confining pressure was kept constant, while the loading ram of the triaxial apparatus was moved onto the specimen with a constant velocity. From Fig. 1.7 it can be seen that, despite the specimen was unloaded by changing the sense of deformation, the axial strain ε_a was not recovered, irreversible strains (plastic deformations) remained. The volumetric strains ε_v showed additional compaction with a hysteresis when reloading.

Soil stiffness is important in geotechnical engineering, as it determines strains and displacements in structures and also the ground. Usually, in calculations stiffness is assumed as constant, though this is not consistent with experimental results, due to the non-linearity for different gradients of the stress-strain curves, as shown in Figures 1.4 and 1.5. The stiffness at point A in Fig. 1.8

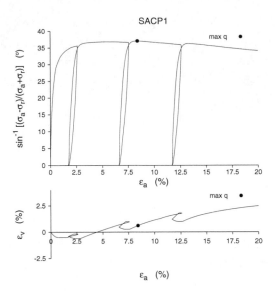

Fig. 1.7: Results of a triaxial test with loading-unloading cycles (test SACP1)

e.g., can be expressed either as tangent or as secant modulus:

$$E_{tan} = \frac{d\sigma'}{d\varepsilon} \quad , \tag{1.1}$$

$$E_{sec} = \frac{\Delta\sigma'}{\Delta\varepsilon} \quad . \tag{1.2}$$

Fig. 1.8 shows the evolution of E_{sec} versus axial strain ε_a, measured in the triaxial tests with different confining pressures and nearly the same initial density (same tests as in Fig. 1.4). The stiffness at small strains is drastically decreasing with increasing strain. The small strain range[6] is defined from $\sim$0.001% to $\sim$1%.

The E_{sec} and E_{tan} are nonlinear with respect to strain (Fig. 1.8) and increase with confining pressure, as shown in Fig. 1.9, where the secant stiffness E_{sec} is plotted against the axial strain for five different confining pressures. E_{sec} decreases nonlineary under loading and increases with confining pressure. The first readings in these tests were around 0.001% of axial strain.

The strains in geotechnical field problems can be small. Hence, conventional triaxial tests are not suitable for the reliable measurement of soil stiffness in small strain ranges. The axial deformation in the triaxial tests, presented in

---

[6]Atkinson, J.: An Introduction to the Mechanics of Soils and Foundations, McGraw-Hill Book Company, London 1993

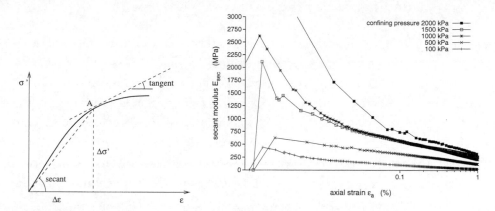

Fig. 1.8: Tangent and secant stiff- Fig. 1.9: Nonlinearity of stiffness, exemplarily shown
ness   on tests on Ottendorf-Okrilla-sand

this study, was measured using internal local displacement transducers. For
the determination of soil stiffness at even smaller strains than ∼0.001%, so-
called dynamic devices such as bender elements are necessary. Fig. 1.10 by
Menzies[7] shows the range of applicability of strain measurements in triaxial
tests.

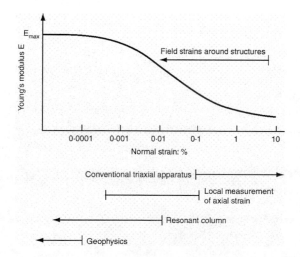

Fig. 1.10: Range of application of stiffness measurement in triaxial tests, after Menzies[7]

[7]Menzies, B.K.: Near-surface site characterisation by ground stiffness profiling using
surface wave geophysics; Instrumentation in Geotechnical Engineering, H.C. Verma Commemorative Volume, Eds. K.R. Saxena and V.M. Sharma, Oxford & IBH Publishing Co.
Pvt. Ltd., New Delhi, Calcultta, 2001, pp. 43-71

1.1.3   Hardening, softening

Loading of granular soil specimens causes deformation and movements of the grains. These displacements represent strains. The force associated with the deformation corresponds to stresses within the body. Stress and strain are related to each other, i.e. by the material behaviour. Stress can be increased nearly without limits if the soil is oedometrically (one-dimensionally) compressed.Whereas in triaxial compression, shearing of the material takes place. Loose soils reach a plateau of deviatoric stress, whereas dense soils form a peak, the curve declines after this peak (*softening*). Continuation of straining leads to a plateau, which coincides for both loose and dense soils (critical state). When the critical state is reached no further volume change takes place. This is difficult to achieve in conventional triaxial testing due to localization of deformations in the triaxial specimen, where in the localized zone a higher void ratio develops than in the remaining part of the specimen, see section 3.3.2.

From Fig. 1.7 it can be seen that reloading can lead to yielding at higher stress (*hardening*). The specimen gets harder due to higher density after each new loading step. The loading history is thus important for the material behaviour.

1.2   Large scale tests

Field tests require extensive measurements of soil properties and monitoring of the investigated problems. In-situ can be measured on existing structures (e.g. long-term observations to assess stability of historic buildings), or the measurements are conducted during the construction process. In the following some cases are described in brief.

1.2.1   Examples from the literature

Retaining walls: v. Wolffersdorff[8] reports on an in-situ test of a sheetpile wall in Hochstetten (Germany). The soil was extensively tested in-situ (4 exploration drillings, 26 soundings, 2 plate load tests) and in laboratory (soil

---

[8]v. Wolffersdorff, P.-A.: Verformungsprognosen für Stützkonstruktionen (Predictions of deformations of retaining structures), Publications of the Institute of Soil and Rock Mechanics, University Fridericiana in Karlsruhe, Vol. 141, 1997

classification, 18 grain size distributions, 17 oedometric compression tests, 9 direct shear tests, 22 triaxial compression tests). A cross section and a plan view of the test site is shown in Fig. 1.11.

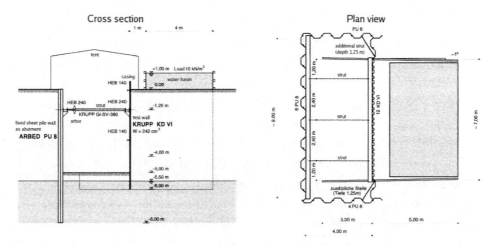

Fig. 1.11: Cross section and plan view of the in-situ sheetpile wall test, taken from v. Wolffersdorff[8]

The sheetpile wall was instrumented with three rows of each 7 ground surface measurement points behind the wall, 22 earth pressure gauges, 6 inclinometer, one temperature sensor. The test was conducted in 8 steps (steps 1-6: excavation, step 7: filling of water basin and step 8: release of struts). The results were processed into many diagrams, which show e.g. bending moments in the sheet piles, earth pressures, forces in the struts, temperature, and displacements.

The earth pressures, bending moments and displacements of this sheetpile wall were predicted by 43 parties in a contest. The calculations comprised non-linear FEM, subgrade modulus methods, empirical methods. The predicted results showed large variation. Therefore, v. Wolffersdorff concludes that, although the soil profile was simple, the behaviour of such a sheetpile wall cannot be grasped with the available calculation methods. One reason for this shortcoming can be found in the inadequate assessment of compaction of the soil which surrounds the wall due to the pile-driving in the calculations.

Extensive measurement data on Berlin walls in cohesive and non-cohesive soils can be found in several publications by Weißenbach[9]. With the help of these

[9]Weißenbach, A.: Reports on Measurement and Evaluations of measurements, in Series of the Department "Baugrund-Grundbau" University of Dortmund, Volumes 3-15 and 17-

measurements, recommendations for design and construction were developed.
A presentation of this measurements surely would go beyond the scope of this
synopsis of recent field measurements. The interested reader is refered to the
publications by Weißenbach.

Foundations: v. Wolffersdorff[8] reports on load tests on footings. The footing, see Fig. 1.12, was loaded with an inclined and excentric 2000 kN hydraulic
jack. During the test following quantities were measured: (i) load P with a
load cell and via recordings of hydraulic pressure of the jack, (ii) load direction
with inclinometer, (iii) displacements with 6 rope sensors and with nivellements, (iv) ground heaving adjacent to the foundation with nivellements, and
(v) variation of anchor lengths of the tripod foundation. The results were
processed into diagrams. The results of this in-situ test were predicted on the
basis of extensive field investigations and back-calculated by several parties
with various constitutive laws and FE-programs. The prediction-competition
found much less echo than the aforementioned predictions for the sheetpile
wall.

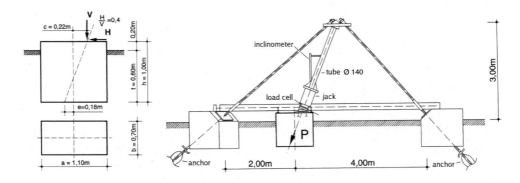

Fig. 1.12: Foundation test in Hochstetten, taken from v. Wolffersdorff[8]

Piles: An extensive measurement program on pile groups for the foundation of a 299 m multistorey building in Frankfurt/Main was conducted by
Holzhäuser.[10] 30 piles were instrumented: 5 of them with load cells at the
pile head, 15 with load cells at the pile tip, and a total of 300 strain gauges

20, 1991-1994

[10]Holzhäuser, J.: Experimentelle und numerische Untersuchungen zum Tragverhalten
von Pfahlgründungen im Fels (Experimental and numerical investigations to the bearing
capacity of pile foundations in rock), Publications of the Institute and Experimental Station
for Geotechnics, Technical University Darmstadt, Vol. 42, 1998

for axial strain measurements. Furthermore, the footing was underside instrumented with 13 contact pressure cells and 4 piezometers. The settlement was monitored with 13 multiple extensometers down to 105 m below ground surface. Alltogether, 100,000 single data were obtained every day. The building was erected on stiff Frankfurt clay and founded in Frankfurt limestone. The measurements showed that the building load is nearly completely transfered by the piles into the limestone.

Pile integrity: In order to detect defects in piles with dynamic pile integrity tests, Plaßmann[11] conducted large scale laboratory tests with 24 various piles in air and in quartz sand. To reduce the influence of errors in the production of the piles, defects were stepwise enlarged on the same pile. The piles were 3.5 m long with a diameter of 0.2 m and were equipped each with 3 wire strain gauges. In a first test the measurements were done on intact piles in air. Subsequently the influence of soil was taken into account. The soil around the piles was installed in 5 layers, which were compacted with a drop-weight, whose height and blow counts were kept constant. The soil installation was controled with drop-penetration tests. The piles were induced by hammer strokes. The induced stress waves are reflected at the pile tip and at changes of the pile cross section. The accuracy of defect position detection depends highly on the impact type and the soil compactness.

1.2.2   Foundation tests by Leussink *et al.*

Another foundation was investigated by Leussink *et al.*[12] They tested foundations 1×1 m and 1.5×1.5 m in a test pit in a laboratory. In the test series the density, the depth of foundation and the influence of soil density, mainly on the contact pressure distribution were investigated. Likewise, settlement measurements were conducted. On consideration of these tests, the focus will be on the load-settlement relations (including bearing capacity), as this will

---

[11]Plaßmann, B.: Zur Optimierung der Meßtechnik und der Auswertemethodik bei Pfahlintegritätsprüfungen (To the optimisation of measurement techniques and evaluation methods for pile integrity tests), Publications of the Institute of Foundation Engineering and Soil Mechanics, Technical University Braunschweig, Vol. 67, 2002

[12]Leussink, H., Blinde, A., Abel, P.-G.: Versuche über die Sohldruckverteilung unter starren Gründungkörpern auf kohäsionslosem Sand (Tests to evaluate contact pressure of rigid foundations on cohesionless sand), Publications of the Institute of Soil and Rock Mechanics, University of Karlsruhe, Vol. 22, 1966

be later compared with own tests at a small scale in section 5.2. Therefore, some of these tests are described here more in detail.

Experimental setup

The relative density of the soil defined as

$$I_n = \frac{n_{max} - n}{n_{max} - n_{min}} \quad , \tag{1.3}$$

was varied in the tests from medium ($I_n = 0.5 - 0.6$) to very high ($I_n \geq 1$). Lower densities were not possible due to the self-weight of the equipment. The soil was compacted in layers of 30-40 cm with a vibrating plate compactor for being dense and very dense, and by hand tamping being medium dense ones. Fig. 1.13 shows a sketch of the experimental setup. As abutment for the jack, the very massive hall ceiling was used. The ground was made of Rheinsand with round and unfractured grains 0-5 mm, e_{min}=0.441 and e_{max}=0.776. The shear parameters of the dense sand were determined as φ'_p=37°, c=25 kN/m^2 from a direct shear test and φ'_p=38°, c=20 kN/m^2 from triaxial tests. Very dense sand specimens could not be produced with laboratory equipment for the element tests.

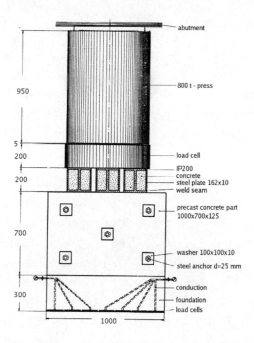

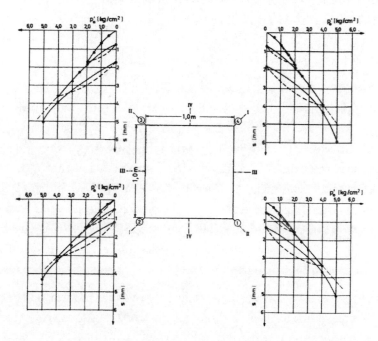

Fig. 1.13: Side view of the experimental setup of the foundation tests by Leussink *et al.*[12] and exemplary result (diagrams adapted) of test series A1/II (1×1 m foundation, $I_n = 0.83 - 0.86$, $t=0$ m), more results in Appendix A.2

Results

Only the load-settlement curves of the experimental results are considered. Fig. 1.13 shows the result of test A1/II (1×1 m foundation, dense soil $I_n = 0.83 - 0.86$, embedment $t = 0$ m). The four diagrams show the stress due to loading p'_0 (without stress due to self-weight of g=0.35 kg/m^2) and the settlement s at the four corners of the foundation. The test was stopped when tilting was observed.

In further tests the embedment depth was varied, while the soil was compacted to approximately same density. Further relevant results and can be found in the Appendix. There are shown the results of test A_1/V with $t = 0.25$ m embedment, test A_1/VI with $t = 0.64$ m. Both footing tests were done in dense soil ($I_n = 083 - 0.86$). More homogeneous compactions can be achieved with higher compaction energy. Therefore, the next test series was done with very dense Rheinsand ($I_n = 1.01 - 1.18$). The results of tests A_1/IX ($t = 0.0$ m, $I_n = 1.18$), $A_1/VIII$ ($t = 0.25$ m, $I_n = 1.01$) and A_1/VII ($t = 0.5$ m, $I_n = 1.15$) are shown in Appendix A.2. There were done more tests with footings 1×1 m which were equipped with more load cells (98 units) and footings 1.5×1.5 m on dry and wet sand. These tests will be not considered in this study for reasons of simplification.

The footing tests presented above were very elaborate. One test required moving of approximately 225 m^3 sand to the test pit. The sand was transported using a belt conveyor to the pit in heaps for 30-40 cm thick layers, which were individually compacted with vibrating plate compactors. The compaction was controlled in 4 levels with a substitution method. If footings were embedded into the ground, the soil was removed by hand at that place after complete filling of the test pit. After each test the pit was emptied. Every load step (+20 t for dense, 10 t for medium dense soil) took 20 minutes time. One test required about 6-8 weeks! The installation of dry sand generates a remarkable formation of dust. Quartz dust is highly noxious.

A repetition of these tests in small scale is shown in section 5.2. The model test results with Soiltron were similar to the ones by Leussink et al.'s. Each of these tests was done in one day!

1.3 Physical models

Due to the complexity of soils and construction processes in-situ soil-structure-interaction can only be realistically grasped in simplified model tests. We have to distinguish between demonstration tests and qualitative experiments. Sometimes, soil-analogue materials are used, such as spheres or cylinders of various diameter.

1.3.1 Typical applications in geotechnical engineering

A physical model is a copy of a prototype situation in small scale. Physical models are often used in combination with numerical simulations in the field of geotechnical engineering. Data from model monitoring can be used for the calibration of numerical models. Whereas a full-scale field monitoring would be favourable, but also very expensive and longsome, as well as difficult to evaluate due to their complexity.

Applications of physical modelling are[13]:

- construction processes (e.g. pile driving),
- cyclic loading effects (e.g. earthquake-induced liquefaction, off-shore structures), or
- transport processes within the soil (e.g. remediation processes),

Physical modelling comprises both small scale models under normal gravity conditions ($1g$ models) and small scale models under centrifuge acceleration (ng models).

When working with small scale models, one has to consider the scaling laws (section 1.3.2), scale effects (section 1.3.3) and limitations of observations (section 3.3.4). The compliance with scaling laws can be proved with dimensional analysis, where all quantities involved in the investigated problem are listed, and dimensionless relations between the relevant quantities are established to express the similarity between model and prototype. The scaling laws have to be adapted for specific problems.

---

[13]Randolph, M.F., House, A.R.: The complementary roles of physical and computational modelling, *International Journal of Physical Modelling in Geotechnics*, 1 (**1**) 2001, pp. 1-8

1.3.2   Mechanical and geometrical similarity

BUCKINGHAM formulated the Π-theorem according to which similarity between model and prototype exists when dimensionless Π products in model and prototype are identical,

$$\Pi_{\text{model}} = \Pi_{\text{prototype}} \quad .$$

By combination of the relevant variables of the problem (Π-variables) one can find functional formulations for the description of the problem.

An example of the use of dimensional analysis is shown on a earth dam (Fig. 1.14), left the prototype and right the geometrically downscaled model.[14] There, the settlement s of the dam in model and prototype may be inspected.

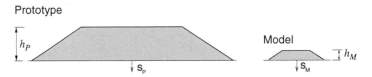

Fig. 1.14: Example: earth dam

The settlement of the dam depends mainly on the height h (m), the bulk density γ (kN/m^3) and the stiffness modulus E_s (kN/m^2):

$$s = f(h, \gamma, E_s) \quad .$$

The dimensionless formulation of the settlement is, e.g.:

$$\frac{s}{h} = f\left(\frac{\gamma h}{E_s}\right) \quad , \text{ or } \qquad s = \int_0^h \frac{\gamma z}{E_s(z)} dz \quad .$$

If stiffness would be proportional to stress, as often assumed, then we could describe stiffness as a linear function of stress $E_s = k\sigma = k\gamma z$, thus, $s = h/k$. From this it follows $s/h = 1/k = \text{const}$ for model and prototype. This would imply that we can use the scaled-down model.

$E_s \sim \sigma$ means that stress ratio vs. strain curves for triaxial tests at different confining pressures would coincide. In reality the normalized stress-strain

---

[14]For simplification reasons gravity is in the beginning neglected. After construction of the dam, gravity is 'switched on'.

curves, and also volumetric strains of a soil at different stress levels do not coincide in the triaxial test. This effect is referred to as *barotropy*, see section 1.1.1. Hence soil is *not* self-similar as by $E_s \sim \gamma z$.

In Chapter 2 will be explained how the pressure dependence of soil mechanical behaviour can be simulated with the similarity in density-dependent and pressure-dependent behaviour.

1.3.3 Problems with physical models, scaling and scaling effects

Parameter	Prototype	1:n model at $1g$	1:n model at ng
Linear dimensions	1	$1/n$	$1/n$
Area	1	$1/n^2$	$1/n^2$
Volume	1	$1/n^3$	$1/n^3$
Stress	1	$1/n$	1
Gravity	1	1	n
particle diameter	1	1	1

Tab. 1.1: Typical scale factors for conventional and centrifuge model testing, n being the scaling factor and g the gravitational acceleration

Table 1.1 gives typical examples for scale factors in conventional and centrifuge model testing. As can be seen, if the geometrical scale is $1 : n$ not all the other quantites are scaled down with the same scale factor. Usually, some Π-variables are allowed to deviate from complete similarity if their influence is negligible in the investigated phenomenon. E.g., the particle size cannot easily be scaled with the same magnitude in the physical models as the other geometrical scales of the linear dimensions of the embedded structures in the model, which is referred to as a *scaling effect*. The linear dimensions are scaled down by the factor $1 : n$, whereas the particle size remains often $1 : 1$ as in the prototype.[15] In the literature can be found several concepts to avoid scale effects. After Ovesen[16] the minimum dimension of structures

---

[15]Note, that effects of particle size cannot be incorporated in the most numerical models, where soil is usually a continuum.

[16]Ovesen, N.K.: Panel discussion to "The use of physical models in design", in *Design Parameters in Geotechnical Engineering*, Proceedings of the VIIth European Conference on Soil Mechanics and Foundation Engineering, Brighton/England 1979, British Geotechnical Society, London, 1980, Vol. 4, pp. 318-323

in the model should be larger than 15 times the mean particle size diameter d_{50}, or after Randolph and House[17] the minimum ratio between structural element size and mean particle size should be 200 to avoid scaling effects. Scale effects develop whenever shear banding occurs. The thickness of shear bands is usually assumed as $5 - 10$ times d_{50}. Tejchmann[18] reports about direct shear tests to assess the wall friction angles between Karlsruhe sand (different layer heights) and steel (rough) and the thickness of the shear zone. The wall friction angle depends on the thickness of the sand layer and the mean grain size, tests with a thickness $h \leq 7.5$ mm of the sand specimen appeared not to be relevant due to material loosening in the contact zones of sand and boundaries (see also Fig. 3.37, page 106). The thickness d_s of the shear zone can be roughly assumed as

$$d_s \gamma_d = \gamma_{cr}(d_s + u_v) \quad , \tag{1.4}$$

with γ_d being the initial dry density of the sand and γ_{cr} being the density at which the sand begins to dilate, and u_v being the maximum vertical displacement.

The boundary of the model container can constrict movements of the soil model particles, and should be, therefore, sufficiently far away from the investigated region. After Andersen[19]: a model container with a diameter which was about 5 times the structural element diameter was found to be too small.

However, geometrical downscaling of the soil particles could change the soil mechanical behaviour in the model. In practice, the same or marginally downscaled particle sizes are used in physical models. And, reasonably, the same soil is used in the model test. The major advantage of Soiltron is that the prototype soil can be used in the $1g$-model, see Chapter 2.

In order to obtain reliable results from small scale model tests, the model has to reproduce similar strength and deformation characteristics as the prototype. In other words, the model has to be as realistical as possible. Several

---

[17]Randolph, M.F., House, A.R.: The complementary roles of physical and computational modelling, *International Journal of Physical Modelling in Geotechnics*, 1 (**1**) 2001, pp. 1-8

[18]Tejchmann, J.: Modelling of shear localisation and autogeneous dynamic effects in granular bodies, Publications of the Institute of Soil and Rock Mechanics, University Fridericiana of Karlsruhe, Vol. 140, 1997

[19]Andersen, K.H.: Panel discussion to "The use of physical models in design", in *Design Parameters in Geotechnical Engineering*, Proceedings of the VIIth European Conference on Soil Mechanics and Foundation Engineering, Brighton/England 1979, British Geotechnical Society, London, 1980, Vol. 4, pp. 315-317

preparation methods were developed to achieve appropriate soil fabric, as e.g. soil layering with subsequent compaction by tapping, or several pluviation methods. This applies also to specimen preparation for the triaxial tests and is described in detail in section 3.1.5. Soil in-situ is usually layered. To model a complete soil profile is nearly impossible, and it is also impractical to draw conclusions from the evaluations. Local inhomogeneities are an important source of errors. Andersen[19] quotes experiments by Rowe and Craig in which 4% loose zones were added (so-called *soft pockets*) in the soil model made of sand (relative density D_r=43%) to evaluate the influence of local inhomogeneities on deformation. The horizontal displacements increased by 50 to 100%.

1.4   Centrifuge models

Centrifuges are often used in physical modelling of geotechnical problems. Different types of centrifuges are being used, (i) beam or (ii) drum and mini drum types.

Centrifuge modelling is very laborious and expensive. Centrifuges are used for the simulation of many geotechnical problems and processes. In the following examples of centrifuges of the world and some application examples are described. Centrifuges are applied e.g. for the simulation or modelling of

- consoildation processes,
- stability problems (e.g. slopes, dams, etc.),
- tunnelling problems,
- foundation simulations (e.g. deep and shallow foundations, piles),
- earthquake actions,
- remediation processes.

The limitations of ng modelling are given in section 1.4.4. Small scale $1g$ modelling has several advantages compared to the centrifuge modelling (see section 1.5).

1.4.1   History

The idea[20] was applied for the first time in the field of geotechnical engineering in the 1930s, independently by P.B. Bucky[21] and G.I. Pokrovsky.[22] In the 1960s at Cambridge University and at Osaka City University trial centrifuges were installed. The research topics were self-weight consolidation and slope stability. Simple data acquisition systems were developed (visual inspection by photographs). Large centrifuges were developed in the 1970s. Data were acquired by computers. Research was mainly in the field of development of pore pressures, effects of stress history, miniature field measurement systems and construction processes, by institutions in the UK, France and Japan. From the 1980s earthquake modelling and subsequent processes as liquefaction and environmental investigations, such as remediation processes, were predominant.

---

[20]This section was inspired by http://www.geotech.cv.titech.ac.jp/~cen-98

[21]Bucky, P.B.: Use of models for the study of mining problems, Technical Publication No. 425, American Institute of Mining and Metallurgical Engineers, 1931, pp. 3-28

Bucky, P.B. and Fentress, A.L.: Application of principles of similitude to design of mine workings, Technical Publication No. 529, American Institute of Mining and Metallurgical Engineers, 1934, pp. 3-20

Bucky, P.B., Solakian, A.G. and Baldin, L.S.: Centrifugal method of testing models, *Civil Engineering*, 5 (**5**) 1935, pp. 287-290

[22]Pokrovsky, G.I. and Fedorov, I.S.: Studies of the soil pressures and soil deformations by means of an centrifuge, In A. Casagrande, P.C. Rutledge and J.D. Watson (eds), Proceedings of the First International Confonference ISSMFE (Harvard), Vol. I, p 70, 1936

1.4.2   Examples of centrifuges

Owner/Highlights	Type	Eff. radius (m)	Payload (kg)	Max. accel. (g)	Capac. (gt)
TU Delft (Netherlands)	beam	1.2	30	300	9

own design and construction, sand pouring machine, 2D loading system, gas supply system, water circulation system, vane apparatus, pile driving hammer, in-flight excavation, wave simulator, simulation of suction pile installation, simulation of pollution behaviour
more: http://dutcgeo.ct.tudelft.nl

ETH Zuerich (Switzerland)	drum	2.2	2,000	440	880

teleobservation via network for educational and operational purposes
more: http://www.igt.ethz.ch

University of California, Davis (USA)	beam	8.5	4,500	53	240

shaking table, gantry robot with changeable manipulator tools, a tool rack for geophysical tests and i.e. pile driving, deformation inspection with high speed cameras, ultrasound or electrical resistivity tomography
more: http://cgm.engr.ucdavis.edu

Univ. of Cambridge (UK)	beam	4.125	1,150	130	150

laser scanner for surface settlement measurements, shaking table
more: http://www-civ.eng.cam.ac.uk

Univ. of Cambridge (UK)	drum	2		500	
	drum	0.8		500	

twin coaxial shafts: one centrifuge (the central turntable, carrying actuators or tools) rotating independently of the other (the ring channel, carrying the soil) allows in-flight construction

Tab. 1.2: Examples of centrifuges, pictures and informations from homepages of the owners

Owner/Highlights	Type	Eff. radius (m)	Payload (kg)	Max. accel. (g)	Capac. (gt)
Ruhr-Uni Bochum (Germ.)	beam	4.125	2,000	250	500
	beam	1.8	400	200	40
	temperature range -196 to +150°C more: http://www.ruhr-uni-bochum.de/zentrifuge				
L.C.P.C. Nantes (France)	beam			200	
	jacks and robots more: http://www.lcpc.fr				
Nat. Univ. of Singapore	beam	2.0	200	200	40
	laser sensors, pile installation tools more: http://www.eng.nus.edu.sg				
Tokyo Institute of Technology (Japan)	beam	2.2			50
	beam	1.2			37.5
	drum	0.6		480	48
	active type shear box and a horizontal–vertical 2D shaker, data transfer with optical rotary joints and slip rings more: http://www.geotech.cv.titech.ac.jp				
US Army Corps of Engineers	beam	6.5	8,000	143	1144
	several climatic conditions are possible more: http://gsl.erdc.usace.army.mil				

Tab. 1.2: Examples of centrifuges (continued)

1.4.3 Examples of model tests in centrifuges

This section is only to deliver a small insight into the variety of centrifuge applications for physical modelling of geotechnical problems. For further information, the reader is refered to the literature.

Modelling of consolidation processes: Due to the small scale of the models, consolidation is performed much faster than in-situ. The validity of similarity rule for the consolidation time t in prototype and small scale model:

$$\frac{t_{\text{prototype}}}{t_{\text{model}}} = \left(\frac{h_{\text{prototype}}}{h_{\text{model}}}\right)^2$$

was shown in experiments by Mikasa *et al.*.[23]

Moo-Young *et al.*[24] report on centrifuge modelling of contaminated seabed sediment, which is capped by a clean soil to reduce the migration of the contaminants from the sediment to the sea. Such caps are often used, but are controversial, since there are many concerns about the migration of the contaminants through the cap. The consolidation induced flow of contaminants was also investigated in centrifuge studies.

Stability problems: The stability of slopes, dams and retaining walls can be assessed, e.g. by increasing the g-level. The height of the prototype structures at collapse can thus be predicted.

Porbaha *et al.*[25] report on centrifuge tests with geotextile reinforced embankment walls and soil slopes with cohesive backfill material. These tests were compared with a numerical model.

Yun and Bransby[26] report on centrifuge modelling of shallow foundations for offshore structures. The foundations were supported by 'skirts' to improve the stability, see Fig. 1.15. For this reason $100g$ centrifuge tests were conducted and compared with raft footings without skirts. The 2D-model was

---

[23]Mikasa, M. and Takada, N. (1973) : Significance of centrifuge model test in soil mechanics, Proc. of 8th ICSMFE, Vol. 1, pp. 273-278.

[24]Moo-Young, H., Myers, T., Tardy, B., Ledbetter, R., Vanadit-Ellis, W. and Kim, T.-H.: Centrifuge simulation of the consolidation characteristics of capped marine sediment beds, *Engineering Geology*, 3-4 (**70**) 2003, pp. 249-258

[25]Porbaha, A., Zhao, A., Kobayashi, M. and Kishida, T.: Upper bound estimate of scaled reinforced soil retaining walls, *Geotextiles and Geomembranes*, 6 (**18**) 2000, pp. 403-413

[26]Yun, G, Bransby, M.F.: Centrifuge modeling of the horizontal capacity of skirted foundations on loose sand, Proceedings ICOF 2003, Dundee

subjected to several load combinations. The horizontal capacity was increased 3-4 times using skirts[27], the occured failure mechanism changed from sliding to rotational.

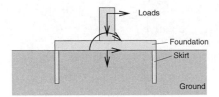

Fig. 1.15: Skirted foundation

Often embankments have to be constructed on soft clay, and a base reinforcement with geotextiles has shown to enhance the stability against spreading of the embankment. Sharma *et al.*[28] report on $40g$ centrifuge tests on clay embankments, reinforced by geotextile and geogrid. The embankment was built in-flight in 20 layers with dry sand using a hopper. Rapid construction of the embankment caused excess pore pressures in the clay foundation and lateral movement of the foundation, geotextile and embankment. An unreinforced embankment failed at approximately 85% of the planned height. However, the geogrid reinforced embankement failed at 95% of the height, and the geotextile reinforced embankment was stable. The lateral movements of the underlying soil was found up to 160 mm in prototype scale after test. The excess pore pressure under the geogrid reinforced embankment was higher than in the geotextile reinforced one, which was attributed to the ineffectiveness of the geogrid reinforcement to reduce the lateral movement of the clay ground, due to the smaller contact surface between soil and geogrid.

Tunnelling problems: In centrifuge tests the face stability and surface deformations can be assessed. Construction processes can be modelled in 2D and 3D.

Schofield[29] reports on tunnel modelling in soft ground. The tunnel was modelled in two scales, 1/75 and 1/125 (concept of modelling of models, see section 1.4.4) as 2D as well as 3D with various depths under surface and

---

[27]Bransby, M.F. and Yun, G.J.: Centrifuge investigation of the horizontal capacity of shallow footings on sand, Proceedings ISOPE 2003, Honolulu, USA, May 2003

[28]Sharma, J.S., Bolton, M.D.: Centrifuge Modelling of an Embankment on Soft Clay Reinforced with a Geogrid, *Geotextiles and Geomembranes*, 1 (**14**) 1996, pp. 1-17

[29]Schofield, A.N.: Cambridge geotechnical centrifuge operations, *Géotechnique*, 3 (**30**) 1980, pp. 227-268

unsupported tunnel lengths. A collapse was induced by reduction of tunnel support pressure. The deformations and pore pressures during tunnel failure were measured. Results of this investigations were stability charts.

To develop a technique for the simulation of sequential segmental tunnel construction in a centrifuge model, 3D tunnel tests were done by Sharma *et al.*[30] with sand in a drum centrifuge. The drum centrifuge was chosen as it is relatively free from boundary effects, see section 1.4.4. The model tunnels were built 'unlined' from paper-covered polystyrene and 'lined' with cylindrical brass foils of different thickness filled with polystyrene. The brass foil, thin enough to create a ground loss of approximately 1-4%, was equipped with strain gauges inside and outside. The length of the tunnel was divided into 4 sections. They were sealed against each other, to allow them to collapse independently, after solving the polystyrene foam inside the brass foil with Inhibisol, an organic solvent (1,1,1-Trichlorethan CH_3CCl_3). The settlements at the ground surface were monitored using a laser profilometer, and visualized.

Kamata *et al.*[31] carried out a series of centrifuge tests on tunnel face stability. Fig. 1.16 shows the general arrangement of the test, as well as a photograph of the tunnel model in a test, where the semi-cylindrical plate which represented

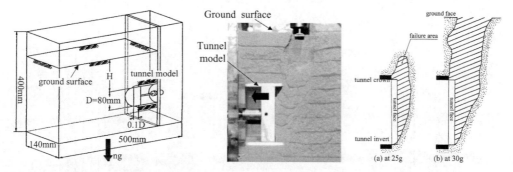

Fig. 1.16: Centrifuge tests for the estimation of tunnel stability, done by Kamata *et al.*: Sketch of the general arrangement, Failure of the not reinforced face at 30*g*, Failure states (no reinforcement); taken from Kamata *et al.*[31]

the tunnel face was pulled back. The overburden/diameter ratio was kept con-

---

[30]Sharma, J.S., Bolton, M.D.: A New Technique for Simulation of Collapse of a Tunnel in a Drum Centrifuge, Technical Report No. CUED/D-SoildTR286, Cambridge University Engineering Department, August 1995

[31]Kamata, H., Mashimo, H.: Centrifuge model test of tunnel face reinforcement by bolting, *Tunnelling and Underground Space Technology*, 2-3 (**18**) 2003, pp. 205-212

stant at $H/D = 1$. Unsaturated Toyoura sand was used, compacted in 2 cm layers. The plate was pulled out when the desired centrifugal acceleration was reached and the tunnel heading face was unsupported. Heading faces without and with reinforcement were tested and visually observed. As reinforcement face bolting of four different bolt lengths in the full section and only the upper or lower half of the section, respectively, was used (Fig. 1.17a). Axial forces and bending moments of the bolts were measured (not described how). The observed bending moment was minimal near the face. The minimal length for bolts was identified as $\geq 0.5D$, as the failure without reinforcement occured in a zone 0.2-0.3D ahead the face. A further series of tests was conducted with vertical bolting from ground surface in front of the tunnel (Fig. 1.17b). Again four different bolt lengths were used at a fixed distance of 0.1D ahead of the tunnel. The effect of this bolting is (verbally) described as face stabilizing when the length of the vertical bolts was approximately equal to the depth of the tunnel invert, shorter lengths led to failure of the tunnel surface. Also the horizontal spacing of the bolts along the cross section of the face for the longest vertical bolts was varied and observed as being favourable at a spacing of 0.1-0.25D. The last series of tests was conducted using forepoling. Again four lengths of bolts and an acrylic plate were used. However, the face stability was not enhanced using forepoling in such bad ground conditions (unsaturated sand).

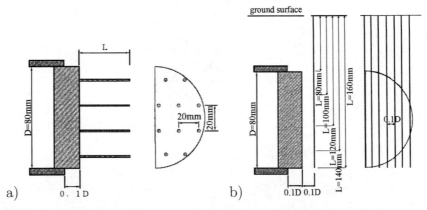

Fig. 1.17: a) Reinforcement of the tunnel face using horizontal bolting, b) Arrangement of the vertical pre-reinforcement bolts, taken from Kamata *et al.*[31]

Foundations: In recent literature it is often reported on deep and shallow foundations and piles, loaded horizontally and/or vertically. Test series with shallow foundations were already described above.

The behaviour of pile groups in clay adjacent to a surcharge load (embankment next to the pile group) was investigated by Bransby and Springman[32], see Fig. 1.18 in centrifuge model tests and back-calculated using 2D-FEM. However, the main focus of this paper was on the description of the finite element calculation. Basics for the centrifuge modelling are given in the Ph.D. theses of both authors. The results of these tests are clearly presented in graphs.

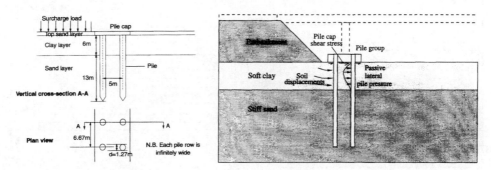

Fig. 1.18: Geometry of the prototype foundation and surcharge loading adjacent to a pile group, taken from Bransby and Springman[32]

Earthquake modelling: Earthquakes can cause damages of soil structures due to liquefaction. There are reports in recent literature about many centrifuge model tests. These are carried out to assess the earthquake hazard of buildings. Major study in this area was VELACS (VErification of Liquefaction Analysis by Centrifuge Studies), a research project sponsored in USA by the Earthquake Hazard Mitigation Program of the National Science Foundation. This study on the effects of earthquake-like forces on a variety of soil models (see Fig. 1.19) at various research institutions aimed at a better understanding of the mechanisms of soil liquefaction and for the acquisition of data for the verification of various analysis methods.

Adalier *et al.*[33] report on four model tests on stone columns in liquifiable silty soils to verify and quantify the possible liquefaction mitigation by application of stone columns. With a first model the response of a pure silt layer was tested, a second model contained 45 stone colums to investigate

---

[32]Bransby, M.F., Springman, S.M.: 3-D Finite Element Modelling of Pile Groups Adjacent to Surcharge Loads, *Computers and Geotechnics*, 4 (**19**) 1996, pp. 301-324

[33]Adalier, K., Elgamal, A., Meneses, J., Baez, J.I.: Stone columns as liquefaction countermeasure in non-plastic silty soils, *Soil Dynamics and Earthquake Engineering*, 7 (**23**) 2003, pp. 571-584

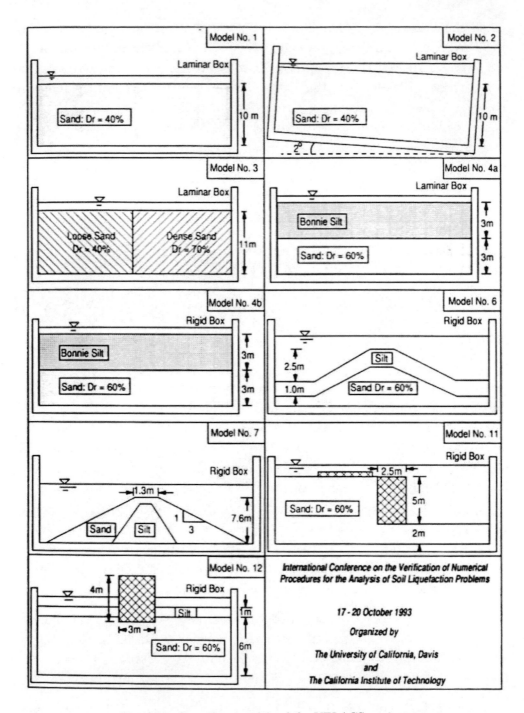

Fig. 1.19: Centrifuge models of the VELACS project,

http://geoinfo.usc.edu/gees/velacs

their impact on lateral stiffness of the silt layer. A third and a fourth models were more or less similar to the first two, but contained an additional footing, simulating a 10-15 store reinforced concrete building, and less stone columns. The stone columns for the construction process were cemented with sugar, the soil model was constructed by air pluviation from a funnel and mild compaction. Upon saturation the cementing effect of the sugar vanished. The response of the stone column equipped models was overall stiffer, settlements were approximately 50% smaller and excess pore pressures could be reduced.

A state-of-the-art of the impact of earthquakes on pile foundations in liquefiable soil was given by Finn and Fujita.[34] During earthquakes piles may shift and settle or crack. Large ground displacements can take place, generating high forces on pile constructions. Piles in endangered areas have to be properly designed, to avoid damage, see Fig. 1.20. In the paper a numerical approach for the calculation is presented and compared with the model test results.

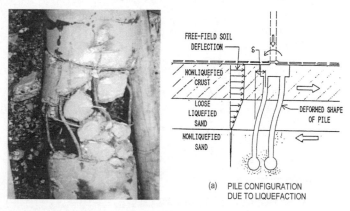

Fig. 1.20: Damage to a pile by 2 m of ground displacement in Niigata earthquake 1964, sketch of the distortion of a pile foundation by lateral moving soil; taken from Finn and Fujita[34]

The following application examples are only mentioned here. The interested reader is refered to the literature.

Geoenvironmental applications: Remediation processes, as the transport of DNAPLs (dense non-aqueous phase liquids) can be faster modelled under higher acceleration, since the fluid velocity is n times higher using ng

---

[34]Finn,W.D.L., Fujita, N.: Piles in liquefiable soils: seismic analysis and design issues, *Soil Dynamics and Earthquake Engineering*, 9-12 (**22**) 2002, pp. 731-742

centrifuge accelleration. Mitchell[35] gives an overview of geoenvironmental applications of centrifuge modelling and summarizes recent publications.

Cold region applications: Recently was reported about tests, that are carried out to investigate the behaviour of frozen soil, such as frost heavals, thaw induced settlements or other ice induced damages, such as scour of icebergs[36], damaging pipelines buried on the sea ground or ice forces on structures.[37]

1.4.4   Advantages/disadvantages of model tests in centrifuges

Reliability of observations: The comparison of the results in the small scale with real field conditions is rather difficult, because field conditions are more complex (soil profile, composition, aging effects, etc.). Often, conclusions have to be drawn from a small number of model tests or field observations. Methods to check the reliability of centrifuge model test results and to detect and quantify experimental errors are: (i) to repeat the tests under similar conditions, or (ii) to repeat the tests in different scales, so-called 'modelling of models'.

Thiam-Soon[38] reports on the concept of 'modelling of models': *Even if similarity is evident in these tests, it seems that only similarity of behavior of different models in a centrifuge has been established and not between the model and uniform g-field prototype behaviors.*

Limitations/boundary effects: Every centrifuge is built for a special range of applications, as the space in the bucket and the payload are limited. Modifications, such as extensions or changes in the simulated construction process, are more difficult to accomplish than in a static model test box.

Due to the limited space in the strongbox of a centrifuge, boundary effects may influence the results of centrifuge tests. To minimize boundary effects

---

[35]Mitchell, R.J.: The eleventh annual R.M. Hardy keynote address, 1997: Centrifugation in geoenvironmental practice and education, *Canadian Geotechnical Journal*, 5 (**35**) 1998, pp. 630-640

[36]Yang, Q.S., Poorooshasb, H.B.: Numerical Modeling of Seabed Ice Scour, *Computers and Geotechnics*, 1 (**21**) 1997, pp. 1-20

[37]Clough, H.F., Vinson, T.S.: Centrifuge model experiments to determine ice forces on vertical cylindrical structures, *Cold Regions Science and Technology*, 3 (**12**) 1986, pp. 245-259

[38]Thiam-Soon, Tan: Two-phase soil study: A. Finite strain consolidation, B. Centrifuge scaling considerations, Technical Report, California Institute of Technology, 1985

due to side friction, lubricants, such as silicone grease, can be applied on the strongbox inner surfaces.

Instrumentation: The instrumentation in centrifuge tests is very sophisticated. All instruments must be of a high accuracy and of small dimensions. They have to withstand high forces, when used inside the modelled soil or under water. Collected data must be transfered during the flight over sliprings to the processor. Processes during the test can be observed with instruments fixed at the arm of the centrifuge (no relative movement to the specimen) or fixed externally (measuring whenever the model passes). Model movements can be measured using deformation gauges (e.g. LVDTs) or remote sensing (e.g. PIV, X-rays).

Centrifugal force and safety for technicians: Centrifuges exhibit high forces. Consider a mass $m = 2,000$ kg of soil rotating at a radius $r = 2$ m at $\omega = 10$ rad/s and a tangential velocity $v = 2$ m $\times$ 10 rad/s $= 20$ m/s. The radially outward acting force (centrifugal force) is $F = m\omega^2 r = 2,000 \times 10^2 \times 2$ m $= 400$ kN ≈ 40 t. Therefore, the arm and the container of the centrifuge have to be very strong. Centrifuges are often situated under the floor of buildings for safety reasons.

Coriolis acceleration: The Coriolis acceleration occurs when a mass is moving in radial direction. Due to the increase of tangential velocity with distance from the center of rotation, a body in radial uniform motion moves spirally outwards. The Coriolis acceleration is $2v\omega$, with ω being the angular velocity and v the velocity of the moving mass. However, the error from the Coriolis effect, expressed as ratio Coriolis/centrifugal acceleration is very small if the velocities are suitably limited. E.g., a high velocity v of 0.1 m/s would induce (assuming a tangential velocity ω=10 rad/s) an error of only 1%.

Accuracy of g-level: The stress distribution in the centrifuge is dependend from centrifugal acceleration and is only accurate at the effective radius R of the centrifuge model, because the g-level varies with depth of the model, see Fig. 1.21. The vertical stress σ_v can be calculated from

$$\sigma_v = \rho g n a R \ ,$$

with ρ being the density, n the geometric reduction scale and $a = t/R$ a factor, defining depth of the radius of centrifugation R under soil surface in the model. This leads, compared to the prototype, where the g-level is constant, to an understress at points above aR and an overstress under aR. To minimize this error big radii and small models should be used.

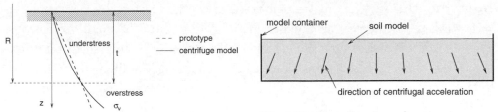

Fig. 1.21: Stress error in the centrifuge

Fig. 1.22: Centrifugal acceleration in a centrifuge model container

Also the direction of g in the centrifuge model container is not realistic, since it is always in radial direction (Fig. 1.22). In the case of small centrifuges the influence can be drastic as can be seen in Fig. 1.23, see e.g. Zeng.[39]

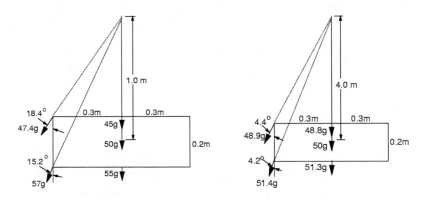

Fig. 1.23: Variation of centrifuge acceleration in models of a 10 m soil layer, taken from Zeng[39]

Zeng[39] made a numerical investigation (with Cam-Clay) of the influence of self-weight in centrifuge models. He found that mainly the horizontal stresses differ enormously in the strongbox of a beam centrifuge from that without centrifugal acceleration, see Fig. 1.24. The difference in horizontal stresses is higher for small radii of centrifugation. The nonuniform horizontal stress distribution causes **lateral spreading**. In order to assess the difference in settlement prognosis using different types of centrifuges, settlement measurements of a 6 m flexible strip footing were compared with numerical simulations

---

[39]Zeng, X.: Benefit of collaboration between centrifuge modeling and numerical modeling, in Proceedings of NSF International Workshop on Earthquake Simulation in Geotechnical Engineering, Cleveland, Ohio (USA), 8-10 November 2001

with various radii of centrifugation. The magnitude was up to 16% overpredicted for small radii.

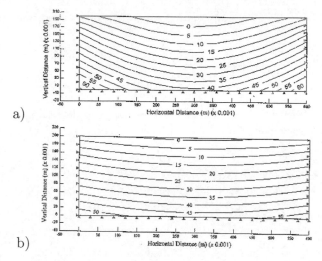

a)

b)

Fig. 1.24: Effective horizontal stresses for a soil layer in a centrifuge model (box width $B = 0.6$ m, centrifugal acceleration $50g$), a) radius of centrifugation $R = 1$ m, b) $R = 4$ m, taken from Zeng[39]

Soil model: The construction of the model depends on the type of soil. Cohesive soil models are, in general, prepared by consolidating soil-water mixtures from the slurry state. Pre-consolidation can be achieved by applying loads onto the model soil under normal gravitation or by self-weight under elevated gravitational conditions. One can also use undisturbed specimens taken in-situ.

The methods to produce soil models from dry non-cohesive soil[40], e.g. sand, are:

1. Hand tamping

   Predefined quantities of sand are spread in the model container and compacted by hand in layers to a predefined thickness. Either the surface can be tamped or the side walls of the container to densify the soil. The main disadvantage of this method is, that it is impossible to achieve homogeneous soil models due to the fact that not all parts of the model are reached equally.

---

[40]For further informations the reader is refered to: Stuit, H.G.: Sand in the geotechnical centrifuge, PhD thesis, Technical University Delft, 1995

2. Fluidisation

Sand is placed into the container and subsequent waterfilled by an upward flow, which is stopped when reaching the surface. Then the sand settles slowly. This method can be used only to produce very loose soil models.

3. Pluviation from hoppers

This method can be categorized into three groups:

— Spot hoppers (single nozzle): The nozzle is moved in regular pattern over the soil model until the desired height or shape is achieved.
— Line hoppers (curtain rainers): Sand is raining like a curtain through a narrow slot, the container is moved horizontally forth and back until the desired model is produced.
— Plane pouring types (sieve rainer): Sand is raining over the whole model container from a bucket through a sieve or a combination of sieves, while keeping the falling height constant.
Note, the falling height must be high enough to assure that the surface is smooth, see Fig. 1.25.

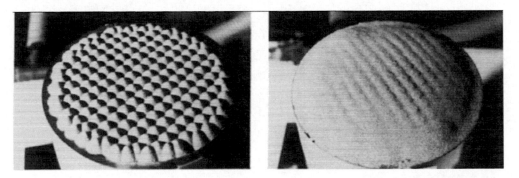

Fig. 1.25: Surface of the soil model using a sieve rainer, a) 50 mm falling height, b) 200 mm falling height, taken from Stuit[40]

Due to the aforementioned Coriolis effects these raining methods are difficult to apply in-flight and, therefore, often the soil models have to be transported into the centrifuge. Due to the transport and provoked shocks, the density, especially of loose models, changes.[41]

Anyway, the surface of these soil models must be smoothed using i.e. vacuum levelers, which suck up excessive sand grains, or using scrapers.

---

[41]Note, that a small shock in model size is like an earthquake in prototype size!

The porosity of the gound model decreases with increasing g-level and can be, as also the overall density of all aforementioned preparation methods, only estimated from the weight divided by the soil model volume. From the measurement of the surface settlement, the change of density during acceleration can be calculated. But, due to the higher g-level in deeper layers of the centrifuge model, the compaction will be higher there. Also the increase of settlement with time has to be considered in centrifuge tests.

Water: To saturate initially dry soil models one can suck de-aired water by means of vacuum from the bottom of the model container into the soil or pluviate dry sand into a water-filled container.

Time scaling in centrifuge modelling is case-dependent, which can be categorized into (i) diffusion type, such as consolidation, and (ii) dynamic type, such as earthquakes. Since the time scale for pore pressure dissipation is scaled to $1/n^2$ and dynamic deformations by $1/n$ (after Dewoolkar *et al.*[42]), one has to use for dynamic investigations a highly viscous fluid, such as silicone oils, glycerin-water solutions or methyl cellulose-water solutions, as pore fluid instead of water to reduce the permeability of the soil model. Dewoolkar[42] report about substitute fluids for centrifuge modelling. The *ideal* substitute fluid should have at least following properties: (i) a density close to that of water, (ii) the same surface tension as water, (iii) should behave like a Newtonian fluid, (iv) it should be polar, to enable the use also with silts, clays and sand, (v) it should have the same viscosity under high gravity conditions as water under $1g$, (vii) the equipment should be inert against it, (viii) the saturation process should be possible in acceptable time, etc. In the aforementioned study, metolose (hydroxypropyl methylcellulose, a water-soluble cellulose ether) was proved as a good model pore fluid for seismic centrifuge tests on saturated cohesionless material, but at the stage of their investigations not fully for cohesive materials.

In-flight construction / In-flight processes: For a realistic simulation, construction processes must be performed in-flight. Typical applications are tunnel boring, pile driving, erection of sheet pile walls or slopes etc.

The in-flight construction and operation is very difficult due to i.e. the Coriolis effect and the high effort in miniaturization. Due to the Coriolis effect e.g. dams cannot be pluviated in layers in-flight, because the grains would not

---

[42]Dewoolkar, M.M., Ko, H.-Y., Stadler, A.T., and Astaneh, S.M.F., A substitute pore fluid for seismic centrifuge modeling, Geotechnical Testing Journal, 3 (**22**), 1999, pp. 196-210

fall radially. Or, if impacts of rocks on structures should be modelled, there a falling weight has to be leaded to the surface of the structure. The friction of the falling weight on the guidance leads to a reduction in velocity.[43]

For the in-flight construction of structural elements a high effort for miniaturization is needed. Construction processes are very complex to replicate in the centrifuge, considering for example a construction pit with sheet pile walls and anchors. Models have to be simplified.

Particle size effect: The grain size cannot be scaled down as the other dimensions of a small-scale model. Consider for example a clay in a small-scale $100g$ centrifuge model with a mean grain diameter of $d_{50,m} = 0.001$ mm would be in prototype scale a sand $d_{50,p} = 0.1$ mm. Clay and sand have different mechanical behaviour, which has to lead to misleading observations in this centrifuge model. Grain size can be expected to play a role in particular if shear bands occur. The average thickness of shear bands is about 10 times the mean particle diameter. The particle size effect on sands is reported[44] to play a minor role, when the relevant dimension of the model D is >15 (better: 30) times of the mean diameter of the model soil d_{50} and the particle shapes and particle hardnesses are similar.

Conclusion: In this section applications of centrifuge modelling were shown. Beside the advantages, such as that the stress is reproduced good in the model, centrifuge modelling has several disadvantages. Therefore, one should not disregard the capabilities of $1g$ modelling.

---

[43]lecture by R. Chikatamarla on GeoDACH seminar 2004, Innsbruck

[44]Modelling in Geotechnics, Course summer term 2003, ETH Zurich - Institute of Geotechnical Engineering, `http://geotec4.ethz.ch/mig/`, Chapter 7

1.5   1*g* versus *ng* models

Small scale modelling under normal gravitational conditions has long been recognized as valuable technique for the investigation of engineering applications and verification of numerical solutions. Due to the stress dependent characteristic of soil mechanical behaviour these tests have limits of applicability.

Some of the advantages and disadvantages of centrifuges also apply for 1*g* model tests (see section 1.4 for details), some do not:

- Limited space in the model container.

  Under 1*g* conditions, extensions or modifications of the model container can be easier done than in the centrifuge bucket, where space and pay load is fixed

- Boundary effects, such as constraints of the side walls of the model box can develop.

- Instrumentation.

  Measurements are easier under 1*g*, since the instrumentation can be simpler adapted to the needs of the tests. Requirements on the measuring devices are lower, they do not have to withstand high forces as during centrifuge flight.

- Soil model.

  The soil model is, usually, for both centrifuge and 1*g* modelling prepared in the same way, but this is easier under normal gravity as compared with e.g. an in-flight construction of a dam. The preparation of the soil model is the main source of errors, cf. the errors in displacement measurement, when adding so-called soft pockets into the soil model, see section 1.3.3, p. 17.

- Particle size effect.

  The prototype soil is usually the same as model soil. Therefore the particles are not scaled. If shear bands occur, they will have the same thickness in the model as in the prototype, approximately 10 times the mean particle diameter. The error caused by this effect is the so-called particle size effect.

1.5.1   Difficulties due to barotropy and pyknotropy

The main problem in 1g models is the high dilatancy and the low stiffness and strength of the soil due to the low pressure. This can be shown on stress-strain curves and volumetric strain curves of e.g. triaxial tests, see Fig. 1.4 or 1.5 on page 4, where results obtained with similar initial density and varying stress level, as well as results with varying initial density and constant confining stress are shown. From the figures it can be seen that with increasing pressure the peak friction angle and the dilatancy are decreasing, and the same occurs with decreasing density: these parameters decrease – a similarity which led to the idea of Soiltron:

With Soiltron it is possible to simulate the in-situ soil behaviour in a small scale laboratory 1g model with the same soil in a looser state.

It is, however, very difficult to work with loose soil, because a loose skeleton collapses with the slightest vibration. Therefore, we developed Soiltron – a soil with loose but stable grain skeleton. This was achieved with a mixture of soil and a granular additive which is soft and light, see Fig. 2.1, p. 49. Soiltron is described in detail in section 2.2.

1.5.2   Initial soil density in the models

Forces and strains are usually measured at the boundary of the specimens or models. Homogeneity of the investigated soil or model is very important. Methods to prepare a soil model were described in detail on pages 33 to 35. In the author's model tests of Chapter 5 the undercompaction method, as e.g. described by Doanh *et al.*[45], was used to prepare soil models. Usually the initial density is obtained from the weight and the dimensions of the specimen/model. With the undercompaction method the density of each installed soil layer is controled and the additional densification due to thereupon installed layers is taken into account, as the layers are compacted to a desired height. The layers are clearly visible in at the boundary of a soil model Fig. 1.26. With this method a rather uniform density is achieved.

---

[45]Doanh, T., Ibraim, E., Matiotti, R.: Undrained instability of very loose Hostun sand in triaxial compression and extension. Part 1: experimental observations, *Mechanics of Cohesive Frictional Materials*, 1 (**2**) 1997, pp. 47-70

[47]The photograph was taken on a model test to evaluate settlements due to shallow tunnelling; test done by M. Mähr, PhD student at the Institute of Geotechnical and Tunnel Engineering, University of Innsbruck

Fig. 1.26: Soil model (Ottendorf-Okrilla sand) visible through a transparent side wall of the model box. The soil model was prepared using the undercompaction method.[47]

Frost and Park[48] report on triaxial specimen preparation using pluviation through air and water, and undercompaction. The soil specimens were saturated with resin and cut into slices for the observation with X-rays, as shown in Figures 1.27 and 1.28, and quantitatively analysed using the gray scales of the images. For the specimen prepared in layers (Fig. 1.27) the scatter in the gray scales is more pronounced than in the air-pluviated ones (Fig. 1.28). Another disadvantage of the undercompaction method is that the specimens are pre-loaded due to compaction.

1.5.3   Measurement of displacements and forces

Displacements can be measured with the help of e.g.

- displacement transducers (mechanical or electrical),
- optical methods, such as PIV (see section 3.1.4), or
- X-ray, possibly with computer tomography.

Displacement transducers measure only locally, whereas with optical methods displacements can be evaluated at many points. Care must be taken not to influence the soil model with stiff measurement devices. In the author's model tests (Chapter 5) displacements were measured using both electrical and mechanical displacement transducers.

---

[48]Frost, J.D., Park, J.-Y.: A critical assessment of the moist tamping technique *Geotechnical Testing Journal*, 1 (**26**) 2003, pp. 57-70

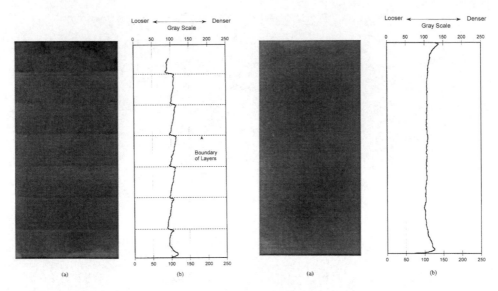

Fig. 1.27: Triaxial specimen prepared with undercompaction method: (a) X-ray image of sectioned specimen; (b) gray scale analysis on X-ray image; taken from Frost and Park[48]

Fig. 1.28: Triaxial specimen prepared with air-pluviation: (a) X-ray image of sectioned specimen; (b) gray scale analysis on X-ray image; taken from Frost and Park[48]

Forces can be measured in model tests with e.g.

- load cells, or
- strain gauges (resistive wire strain).

In the model tests of Chapter 5 the force was measured with a load cell. The load cell was calibrated by applying metal weights.

The contact stress between soil mass and engineering structures can be measured, e.g., with oil-filled pressure cells. Important for the use of these devices is (i) that their stiffness should be in the same range as the stiffness of the surrounding material, (ii) it should be sufficiently robust to avoid damage during installation, (iii) it should be resistant against corrosion and damaging environmental influences, (iv) temperature compensated, and (v) the signal should be stable. Anyway, pressure gauges were not used in the model tests of Chapter 5.

1.5.4   Examples of 1g model tests from the literature

Foundations: Hettler[49] investigated the influence of monotonic and cyclic loadings on rigid and elastic foundations in sand. He investigated the influence of force, embedment, radius and soil density on the load-settlement characteristics of several footings (and also piles) from the literature. With own experiments he summarised his conclusions in simple model laws which comprised only two soil-independent constants and derived a solution for the settlement of rigid monotonic loaded footings. For cyclic loaded footings three cases were observed: shake-down (decrease of settlements with load cycles), incremental collapse (linear increase of settlement with load cycles, therefore, loss of serviceability due to too large deformations) and alternating deformations.

Retaining walls: Model tests were done by Al-Akel[50] to obtain hints for the design of rigid retaining walls. For simplification the force due to earth pressure on the back of the retaining wall was idealised with a horizontal force at 1/3 of the wall height, see Fig. 1.29. The force was exerted with a hydraulic jack. The model wall was 0.10 m thick, 1 m wide and 1.6 m high (with an embedment in the soil of 1.2 m). The wall was equipped with load cells on both sides. The ground surface was monitored using several strain gauges. The wall was erected on dry sand and back-filled with a sand rainer. The wall was strain-controlled loaded with the hydraulic jack. Results of the tests are, e.g., that the density of the soil has minor influence on the soil-wall friction angle and that the supporting earth-pressure had no upper limit.

Another model test series for the determination of earth pressures was done by Arnold.[51] His investigations were done on buttress walls. A 97 cm wide model wall with 94 cm height and footing length of 64 cm was placed onto a sand layer into a test pit, Fig. 1.30a. The ratio of heel/height length was variable. Deformations of the wall were measured in 4 horizontal positions and

---

[49]Hettler, A.: Verschiebungen starrer und elastischer Gründungskörper in Sand bei monotoner und zyklischer Belastung (Displacements of rigid and elastic foundations in sand under monotonic and cyclic loading), Publications of the Institute of Soil and Rock Mechanics, University of Karlsruhe, Vol. 90, 1981

[50]Al-Akel, S.: Modellversuche an einer im Boden eingespannten Wand (Model tests on a wall fixed in soil), Ohde-Colloquium 2001, in Pubblications of the Institute for Geotechnics, Technical University of Dresden, Vol. 9, 2001

[51]Arnold, A.: Modellversuche zum Erddruck auf Winkelstützwände (Model tests to the earth-pressure on buttress), Ohde-Colloquium 2001, in Pubblications of the Institute for Geotechnics, Technical University of Dresden, Vol. 9, 2001

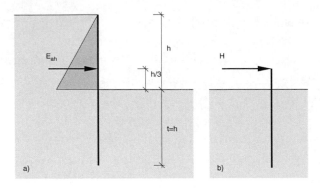

Fig. 1.29: Investigated problem by Al-Akel[50]: a) prototype, b) simplificated system

vertical at the wall head, Fig. 1.30d. The earth-pressure was measured with
18 pressure gauges. The back-fill was placed in layers, and after reaching the
desired height it was loaded with a uniform pressure. The tests revealed that
the earthpressure is underestimated by the common standards (e.g. German
standard DIN 4085 or DIN V 4085-100). The bending moment at the toe of
the wall is reliably predicted by the standards.

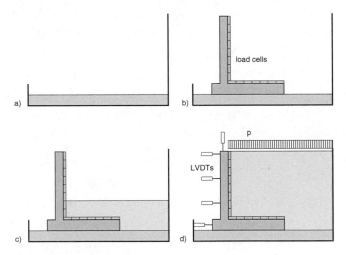

Fig. 1.30: Buttress wall of Arnold[51]: a-d) procedures of the tests

Piles: 29 model piles were tested by Hettler[52] in Karlsruhe sand. He used
model piles made of steel, aluminium, wood, acrylic glass, soft polyurethane
and polyethylene with lengths between 9.6 and 113 cm and diameters from 1.1

[52]Hettler, A.:Horizontal belastete Pfähle mit nichtlinearer Bettung in körnigen Böden
(Horizontal loaded piles with non-linear bedding in non-cohesive soils), Publications of the
Institute of Soil and Rock Mechanics, University of Karlsruhe, Vol. 102, 1986

to 15.9 cm. The piles were horizontally loaded at their heads in order to find a realistic bedding approach. The tests revealed that the surface conditions have a major influence on the pile deformations (displacements of smooth piles were 3 times higher than for rough piles).

Tunnelling: Another series of model test was made in Karlsruhe to assess stability of shallow tunnels near the face (3D case) and far from the face (2D case). These tests were done by Mélix[53] in 6 different soil mixtures of medium sand, clay and water. The model box was made of perspex, the side walls were screwed together, so that they could be dismounted after the test and photographs could be taken. The soil was built in layers by hand. Pluviation was not possible with the cohesive material. The surface of each layer was loaded with 10 kN/m² using a flat jack. Subsequently each layer was smoothed. The tunnel was driven with a pipe, where the soil at the heading face was removed using a cutter. Fig. 1.31 shows the experimental setup. During the driving procedure the ground surface was visible, so that possible

a)   b)   c)

Fig. 1.31: Experimental setup for the tunnel model tests of Mélix[53]: a) flat jack, b) tunnel before driving, c) cutter head

daylight collapses or fissures could be detected. The collapse was effected by increase of the surface load (via the flat jack) or by decrease of inner pressure (via decrease of pressure in a rubber tube inside the tunnel). The surface and the cross section wer monitored during the tests. The tunnel was uncovered after the tests in several vertical cuts. Moistening the soil created a sufficient cohesion so that vertical cuts were possible. The deformations were made

---

[53]Mélix, P.: Modellversuche und Berechnungen zur Standsicherheit oberflächennaher Tunnel (Model tests and calculations to the stability of shallow tunnels), Publications of the Institute of Soil and Rock Mechanics, University of Karlsruhe, Vol. 103, 1987

visible with layers of dark sand. The tests showed that for overburden to
tunnel diameter ratios $H/D < 2$ collapse is the design loading case of such
tunnels. The experimental results were compared with several theoretical
approaches.

Slope stability: In the aforementioned work by Mélix[53] some tests to assess
the mechanical properties of the sand mixtures used in the tunnel models
were conducted. Particularly the cohesion was difficult to assess with common
laboratory experiments. Therefore, the cohesion was assessed by testing the
stability of a vertical slope, see Fig. 1.32. From the measured collapse load
can be obtained the cohesion.

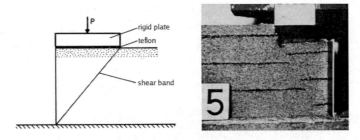

Fig. 1.32: Sketch to the experimental setup of Mélix[53] and cross section after collaps

Wooden piles: Jovanović[54] describes detailed $1g$ model tests on wooden
piles in soft ground, which were conducted in order to explain the compli-
cated bearing behaviour of historical foundations in typical loading situations.
Timber piles were used until the early 20th century primarily for the foundation of buildings in grounds with high ground water level. They usually
consist of vertical piles or a combination of horizontal and vertical piles. To
prevent rotting, the piles have to be permanently under water. The problem
was divided into three groups of model laws:

a) Model laws regarding the foundation construction (geometry, load distribution and stiffness ratios)

The geometrical similarity is given when the ratios of all linear dimensions
are equal. The same applies for the loads. Mechanical similarity is given if
the ratio of stiffnesses of timber and soil in model and prototype are equal.
For this reason the model piles were made of balsa and the soil was assumed

[54]Jovanović, M.: Historische Holzgründungen – Tragverhalten in weichem Untergrund
(Historical timber foundations – Bearing capacity in soft ground), PhD thesis, Publications
of the Institute of Soil and Rock Mechanics, University of Karlsruhe, Vol. 153, 2002

as self-similar, assuming linearity of stiffness, after TERZAGHI's consolidation law

$$E_s = \frac{1 + e_0}{C_c}\sigma_1' = \frac{1 + e_0}{C_c}\gamma h \; . \tag{1.5}$$

b) Model laws regarding the overall stability (reproduction of soil strength and the loads on the foundation for drained and undrained ultimate state, reproduction of time)

Stability was investigated for two cases: short-term and long-term stability. The mechanical similarity in consolidation of the soil model and in the loads was given by the geometrical similarity. For the investigation of long-term stability, the critical effective angles of friction φ_s in model and prototype were assumed to be identical. The most critical point in the investigations was the time scale (load steps and loading velocity). Here, the time scale was equal to the square of the geometric scale, if the boundary effects (side drainage, etc.) are neglected. However, it is quite difficult to distinguish between drained and undrained conditions for the historical construction process.

c) Model laws regarding deformations (reproduction of effective stresses in the soil, stiffness modulus, compression of the soil and settlements)

Effective stresses and stiffnesses given due to the geometrical similarity, and the use of (soft) balsa wood led to mechanical similarity for the structure. Further, Jovanović describes why mechanical similarity holds also for strains in model and prototype. The effective stresses in model and prototype can be made dimensionless by

$$\left(\frac{\sigma_{ij}'(\varepsilon_{ij})}{\gamma l}\right)_p = \left(\frac{\sigma_{ij}'(\varepsilon_{ij})}{\gamma l}\right)_m \; . \tag{1.6}$$

This applies because the stress $\sigma_{ij}'(\varepsilon_{ij})$ decreases linearly with depth, the specific weight is almost equal in model and prototype, and the linear dimensions are reduced $1{:}n$. Due to the geometrical reduction of (i) the soil stiffness (because of the lower stress), and (ii) the stiffness of the timber piles (because balsa was used), the strains in the model are also the same as in the prototype for the structure.

$$\left(\frac{\sigma_1'}{E_s}\right)_p = \left(\frac{\sigma_1'/n}{E_s/n}\right)_p = \left(\frac{\sigma_1'}{E_s}\right)_m \tag{1.7}$$

The settlements measured in the model can be converted into prototype dimensions, because all significant geometrical scales were similar. However, the model is only valid for settlements during consolidation, not for secondary settlements (creep), because the model time scale in this phase is 1:1.

The soil model was built in 78 layers of lacustrine clay, each poured through a polystyrene platen equipped with holes and small funnels, so that a clay suspension was sedimented. Two layers were built per day with a sedimentation time of about 5 hours, so the overall time was about 5 to 9 weeks. An almost uniform falling height of the particles in suspension was accomplished by raised sidewalls. The primary consolidation time (graphically determined after TAYLOR) due to self-weight had to be awaited before the installation of the foundations.

The foundation consisted of vertical timber piles (installed by continuous pressure by hand, not with a simulated ram, to avoid shakes), and for some tests additionally of a grid of horizontal timber beams, for one test the foundation consisted of horizontal beams only. The foundations were subjected to loads in steps in order to simulate the construction process of real buildings. Further, the foundations were subjected to several endangering situations, such as too fast loading (no time to consolidate), excavation of a pit with sheet pile wall close to the loaded foundation, or piling close to the foundation.

Pipe driving: The driving force and the ground movements during pipe driving operations in a small scale model were measured by Salomo.[55] The soil model was prepared with a sand rainer in a $2.4 \times 1.0 \times 1.0$ m model box. The model tunnel (outer diameter 10 cm) was equipped with 7 different cutting edges. The pipe was driven into the soil by strain-control. During the pipe driving operations the surface settlements, displacements, reclaimed sand and the driving force were measured. Forces were measured behind the cutting edge and at the pipe end. 74 tests are documented with varying types of cutting edges, overburden, advance length, driving velocities and model soil compactness. With the model tests a failure mechanism was identified, which is based on a monolithic, passive fracture pattern on a spatial failure plane. The 'optimal' angle of the cutting edge was obtained as $\alpha_{opt} = 45° - \varphi/2$. Furthermore, he concluded that both subsidence cavity and uplift zone can be described with a GAUSSian distribution. The maximum heavings were

---

[55]Salomo, K.-P.: Pressenkräfte und Bodenverformungen beim Rohrvortieb (Driving forces and ground movements during pipe driving operations), Reports of the Geotechnical Engineering Institute, Technical University Berlin, Vol. 6, 1980

observed directly over the cutting edge, the limit value of settlements were observed in a distance 20 to 25 times the pipe diameter behind the cutting edge.

Anchors: The bearing capacity of anchors in sand was investigated by Mayer[56] in model tests under laboratory conditions. Due to the relative large dimensions of the model box and anchors (approximately 1 m) the tests were conducted in a test pit. The model soil was pluviated into the model box. The grout was installed with a pneumatic injection device to assure constant injection pressure. To avoid spontaneous lifting of the injection device, the tubing was equipped with brakes. The anchors were pulled strain-controlled with a hydraulic cylinder. Deformations were measured with resistive wire gauges, earth-pressures in the soil body close to the grout body were measured with miniature pressure gauges, and displacements of the soil body with extensometers (see Fig. 1.33). Results of this study were e.g. definition of the minimum grout pressure of 10 bar and the conclusion that due to expansion of the grout the resistance was intensified.

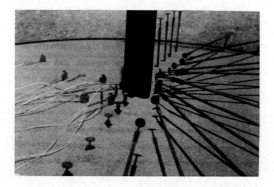

Fig. 1.33: Complex instrumentations in 1g-model tests with anchors by Mayer[56]

In section 1.5.3 measurements in models were described. The complexity of instrumentations in 1g-model tests is visible in Fig. 1.33, where the extensometers and earth-pressure gauges of the anchor tests of Mayer[56] are shown.

Sometimes, the experimental setup of model tests can be very complex, e.g. the one by Jovanović shown in Fig. 1.34 which is described on page 44.

---

[56]Mayer, G.: Untersuchungen zum Tagverhalten von Verpreßankern in Sand (Investigations to the bearing capacity of injection anchors in sand), Reports of the Geotechnical Engineering Institute, Technical University Berlin, Vol. 12, 1983

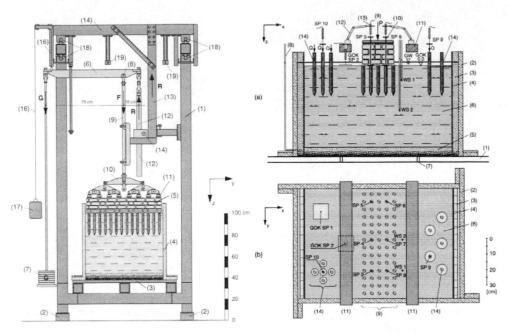

Fig. 1.34: Complex experimental setup for 1g-model tests, e.g timber foundation tests of Jovanović, details to the figures are beyond the scope of this overview, the interested reader is refered to Jovanović's publication

In this section the application of 1g-model tests was shown briefly on some random chosen examples from recent literature. The range of applications is of sheer enormity, and the conclusions from these tests are important for the engineering community. It was shown, that these conclusions can be made under consideration of simple model laws and without the use of centrifuges.

Soiltron as an adequate soil material for 1g-models can offer a new approach to small-scale model tests, as will be shown in the following Chapters.

Chapter 2

Soiltron

-tron: *common greek suffix, e.g. electron is an amber which attracts a feather when rubbed in a woolen drapery. In the style of electron the words positron, neutron, synchrotron, and many more were coined.*

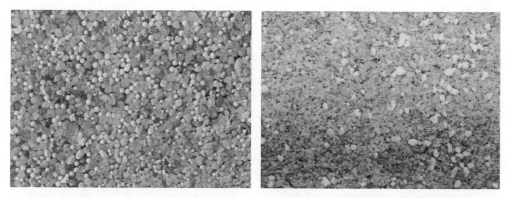

Fig. 2.1: Soiltron; left: Soiltron 1 – Mixture of Ottendorf-Okrilla sand and polystyrene spheres, right: Soiltron 2 – Mixture of Ottendorf-Okrilla sand and perlite grains

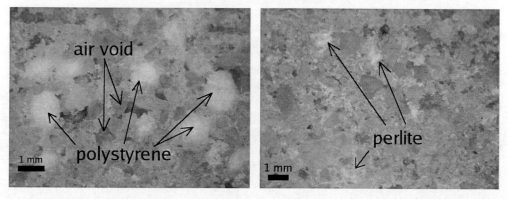

Fig. 2.2: Soiltron embedded in resin (microscope: 12×magnified); left: Soiltron 1, right: Soiltron 2

2.1　Idea

As shown in section 1.1.1, the behaviour of granular soils depends on pressure and density. With decreasing density the peak friction angle decreases, and the volumetric behaviour is more contractant, provided the pressure is constant. The same effect can be observed for increasing pressure, provided the density is constant. The idea is to simulate the soil behaviour at field conditions (higher pressure due to great depth compared to the laboratory model) in a small scale model (lower pressure due to low depth) with a loose soil.

But what is *dense* and *loose*? The relative density

$$D = \frac{e_{\max} - e}{e_{\max} - e_{\min}} \tag{2.1}$$

is acquired in standard compaction tests.

With equ. (2.1) from German Standard DIN 18 126 soil is defined as being loose for $D = 0 \ldots 0.333$, medium dense for $D = 0.333 \ldots 0.667$ and dense for $D = 0.667 \ldots 1$.

The densification of a soil specimen due to the increase of the pressure, here the mean effective stress $p' = (\sigma'_a + 2\sigma'_r)/3$ in triaxial compression, corresponds to a reduction of the void ratio e, cf. Fig. 2.13. The p'-e-relation depends on the initial void ratio e_0, that prevails at $p' = p'_0$. The indices c and d indicate pressure-dependent minimum and maximum void ratios.

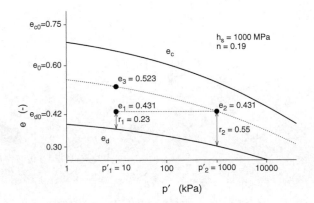

Fig. 2.3: Pressure-dependent void ratios and density factors r_e by equations (2.4) and (2.2)

Taking into account the stress dependence of the void ratio, the relative density of a soil can be expressed as:[1]

$$I_d = \frac{e_c - e}{e_c - e_d} \quad \rightsquigarrow \quad r_e = 1 - I_d = \frac{e - e_d}{e_c - e_d} \quad . \tag{2.2}$$

The mechanical behaviour of the model is the same if r_e is the same as in the prototype. As can be seen from Fig. 2.3, for a certain void ratio $e_1 = e_2$, the relative densities r_1 and r_2 depend on the mean effective stress. Therefore it is often said that a dense soil with void ratio e_1 at the mean effective stress p'_1 (Fig. 2.3) behaves like a loose soil of the same density $(e_2 = e_1)$ under a higher mean effective stress p'_2. The stress and density dependence of material behaviour (barotropy and pyknotropy) was already described in section 1.1.1. The void ratios e_i, e_c, e_d decrease with increasing stress p' according to

$$e = e_0 \exp\left[-\left(\frac{3p_s}{h_s}\right)^n\right] \quad . \tag{2.3}$$

Soiltron has void ratio e_3 at the mean effective stress p'_1. Aim of this study is to demonstrate that a similarity exists between the mechanical behaviour of Soiltron which has the void ratio e_3 at p'_1 in small scale $1g$-models and the prototype soil (in-situ) with e_2 at p'_2 exists.

To achieve the same relative density r_e in the model as in the prototype, the grain skeleton has to be loose at low pressures. However, soil with low density is collapsible and can be stabilized with soft additives (artificial air pores). In this thesis two materials — polystyrene and perlite — were used. The additives influence the mechanical behaviour in a way that the new soil behaves softer. The action of the soft inlets can very well be modelled by means of hypoplasticity: The granular hardness h_s of Soiltron is by circa the geometrical scaling factor n lesser for Soiltron than for the sand which was used in this survey. The similarity in the mechanical behaviour can be utilised to simulate barotropy with pyknotropy.

---

[1]Gudehus, G.: A comprehensive constitutive equation for granular material, *Soils & Foundations*, 1 (**36**) 1996, pp. 1-12

2.2 Composition of Ottendorf-Okrilla sand and associated Soiltron

2.2.1 Ottendorf-Okrilla sand

The used sand was the commercially available quartz "SILIGRAN" of the company Euroquarz[2] from the deposit Ottendorf-Okrilla in south-east Saxonia, Germany (Fig. 2.4). It was multiply washed, dried and sieved by the producer.[3]

It was available in the mineral size fractions 0.1-0.5 mm, 0.5-1.0 mm and 1.0-2.0 mm. The grains are fractured and vitreous (see Fig. 2.4). Several mineral size fractions were mixed together in the ratio 2 parts by weight of fraction 0.1-0.5 mm : 2 parts 0.5-0.1 mm : 1 part 1.0-2.0 mm.

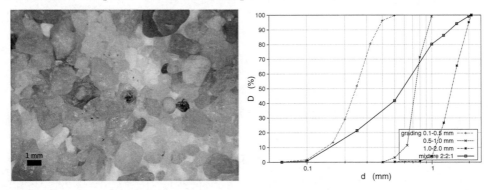

Fig. 2.4: Ottendorf-Okrilla sand: grains under microscope and grain size distribution of the single mineral size fractions and the mixture 2:2:1

---

[2]Euroquarz GmbH, D-01458 Ottendorf-Okrilla, Würschnitzer Straße, Tel.: +49/35205/527-0, Fax: -12

[3]The deposit is located in an about 4 km wide urstromtal, the so-called *Senftenberger Elblauf*. The formation of this deposit began in the Upper Tertiary (11-5.5 milion years ago) and the following ice ages. It has now a thickness of about 35 m.

Grain sizes	d_{10}	0.19 mm
	d_{30}	0.30 mm
	d_{60}	0.69 mm
Uniformity coefficient	$U = d_{60}/d_{10}$	3.6
Coefficient of curvature	$C = (d_{30})^2/(d_{10} \cdot d_{60})$	0.7
Mean grain size	d_{50}	0.58 mm
Particle shapes		angular to subangular
Mineral composition		>97% quartz
Specific weight	γ_s	2.635 g/cm^3
Max. void ratio	e_{max}	0.75
Min. void ratio	e_{min}	0.42

Tab. 2.1: Properties of Ottendorf-Okrilla sand

2.2.2 Types of Soiltron

Two types of Soiltron were investigated in this study. The *working titles* were:

Soiltron 1 which is a mixture of Ottendorf-Okrilla sand and expanded polystyrene (EPS) spheres ∅ 0.1-1.6 mm. Three different ratios of mixture were produced for the investigation of the mechanical behaviour in triaxial tests:

a) **Soiltron 1L:** 0.05 ml EPS per 1 g sand (≈ 1/2000 g PS per 1 g sand)
b) **Soiltron 1:** 0.10 ml EPS per 1 g sand (≈ 1/1000 g PS per 1 g sand)
c) **Soiltron 1H:** 0.15 ml EPS per 1 g sand (≈ 1.5/1000 g PS per 1 g sand)

Soiltron 2 which is a mixture of Ottendorf-Okrilla sand and perlite grains 0.1-1.6 mm in three different ratios,

a) **Soiltron 2L:** 0.0067 g perlite per 1 g sand (≈ 1/150 g perlite per 1 g sand)
b) **Soiltron 2:** 0.0133 g perlite per 1 g sand (≈ 1/75 g perlite per 1 g sand)
c) **Soiltron 2H:** 0.020 g perlite per 1 g sand (≈ 1/50 g perlite per 1 g sand)

The possibility of the mixture was specified in a first step by visual inspection, see Figures 2.5 and 2.6.

Fig. 2.5: Surface of mixtures of Ottendorf-Okrilla sand with polystyrene (see light-coloured spheres): Soiltron 1L, Soiltron 1, Soiltron 1H (from left to right)

Fig. 2.6: Surface of mixtures of Ottendorf-Okrilla sand with perlite: Soiltron 2L, Soiltron 2, Soiltron 2H (from left to right)

2.2.3   Additives

2.2.3.1   Expanded polystyrene (EPS)

The expanded polystyrene (EPS) beads which were used in this study were obtained from the company Röfix AG.[4] These beads are usually used by this company in thermal insulation plasters or light-weight concretes. The diameters of the beads were in the range from 1.0-1.6 mm, with a density of $\rho \approx 10\text{-}20$ kg/m^3 ($\approx 0.01\text{-}0.02$ kN/m^3).

Properties of expanded polystyrene[5]:

- EPS does not rot, is not water soluble,
- is health safe (cf. use for food packings),
- is resistant to e.g. salt solutions, alcohols and acids,
- is soluble in e.g. anhydrous acid, organic solvents and oil,
- the water absorption when submerged 28 days is less than 3 Vol.-%.

Note that the mechanical properties of polystyrene are usually determined in tests on block-moulded EPS. These properties are due to bonding forces between the EPS-beads different from that of the unbounded beads which were

---

[4]Röfix AG, Badstraße 23, A-6832 Röthis, Tel.: +43/5522/41646-0, Fax: -106

[5]Informations from BASF product sheets, http://www.basf.com

used for Soiltron. Extruded polystyrene exhibits low mechanical strength and low stiffness. In multistage oedometric compression tests on soil and Soiltron the time-settlement behaviour was observed, see Fig. 2.11. Soiltron 1 shows a viscous part in the settlements, which can be attributed to the polystyrene beads. The settlements decayed in the various load stages after approximately 1 hour. Therefore, the model tests conducted with Soiltron 1 should be done not to fast, so that the creep settlements come to an end.

When dealing with EPS or other synthetic materials one has to distinguish between slow and rapid deformations, since they are rate dependent materials and tend to creep. Athanasopoulos *et al.*[6] report on experimental investigations of the dynamic properties of *block moulded* EPS geofoams using triaxial tests. Clearly, the mechanical properties of EPS-blocks are different from that of EPS beads, but the investigations of Athanasopoulos *et al.* give an idea of the behaviour of EPS and are used for comparison.

Fig. 2.7 shows a typical result of an unconfined monotonic compression test on EPS geofoam under rapid (10%/min) and slow (45%/year) loading.

Athanasopoulos *et al.* made triaxial tests with various confining pressures (σ'_r=10-60 kPa) and a constant strain rate of 3.3%/min, see Fig. 2.9. As can be seen in that Figure, the elastic modulus of EPS block material is very small.

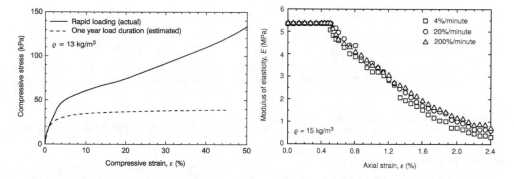

Fig. 2.7: Triaxial compression test on block moulded EPS geofoam (left), Young's modulus of EPS geofoam under different strain rates, taken from Athanasopoulos *et al.*[6]

In Fig. 2.8 are shown the results of drained isotropic consolidated triaxial tests on polystyrene spheres ($\varnothing \leq 2$ mm) and perlite grains ($d = 0.1$-2 mm). The

[6]Athanasopoulos, G.A., Pelekis, P.C., Xenaki, V.C.: Dynamic properties of EPS geofoam: An experimental investigation, *Geosynthetics International*, 3 (**6**) 1999, pp. 171-194

test on the polystyrene spheres was done at a confining pressure of 20 kPa, the one on perlite at 100 kPa. Both materials exhibited a high density in the consolidation phase before the test.

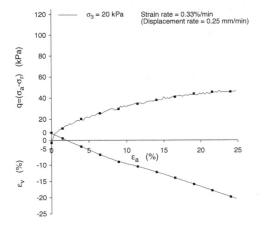 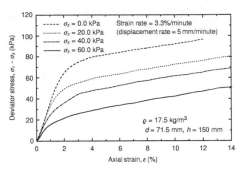

Fig. 2.8: Results of triaxial tests on polystyrene (tested at confining pressure $\sigma'_r = 20$ kPa), deviatoric stress and volumetric strain vs. axial strain

Fig. 2.9: Results of rapid triaxial tests on block moulded EPS geofoam, tested at varying confining pressures $\sigma'_3 = 0$ to 60 kPa, taken from Athanasopoulos *et al.*[6]

2.2.3.2 Perlite

Perlite is an amorphous glass mineral. It is of volcanic origin and expanded up to approximately 20 times its original volume. Typical expanded perlite bulk density is $\rho \approx 30\text{-}150$ kg/m^3 ($\approx 0.03\text{-}0.15$ kN/m^3). There are various types of perlite available in trade: raw (untreated) perlite, and coated with silicon or bitumen. The coating prevents water-absorption. For Soiltron untreated perlite was used.

Perlite is highly compressible and crushable. Linemann *et al.*[7] investigated the degradation of perlite in dependence of pneumatic conveying cycles. The particle size distributions after 1, 20 and 40 conveying cycles are shown in Fig. 2.10. Therefore, in the triaxial tests and the model tests with Soiltron 2 (prepared with perlite) the model material was used only once.

---

[7]Linemann, R., Runge, J., Sommerfeld, M., Weißgüttel: Densification of bulk materials in process engineering, in *Advances in Geotechnical Engineering and Tunnelling* Vol. 3, Kolymbas, D. & Fellin, W. (ed.), A.A.Balkema, Rotterdam, 2000

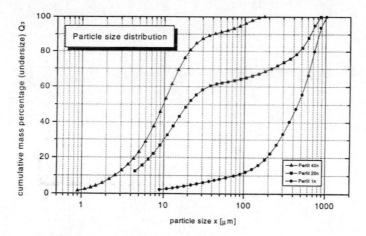

Fig. 2.10: Particle size distribution of perlite in dependence of the number of pneumatic conveying cycles, taken from Linemann et al.[7]

In multistage oedometric compression at various pressures Soiltron 2 shows no viscous behaviour, see Fig. 2.12. The settlements come to an end after approximately 1 hour. Therefore, the model tests with Soiltron 2 should be conducted as mentioned in section 2.2.3.1 in appropriately chosen times.

2.2.3.3   Oedometric compression of Soiltron

Figures 2.11 and 2.12 show the results of multistage oedometric compression tests on Soiltron at various pressures. As can be seen the settlements come after approximately 1 hour to an end.

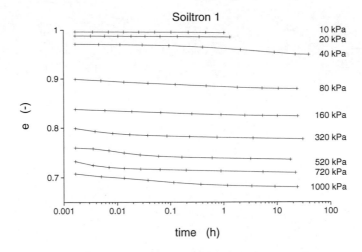

Fig. 2.11: Oedometric test results: time-settlement curves of Soiltron 1 at various pressures

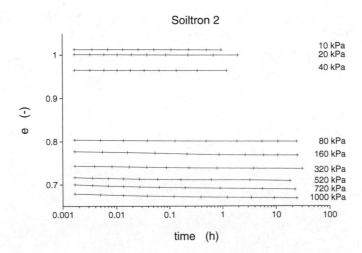

Fig. 2.12: Oedometric test results: time-settlement curves of Soiltron 2 at various pressures

2.3 Constitutive parameters

Constitutive laws describe the mechanical behaviour of materials, such as soils. They are relations between stresses and strains. For soil, simple and complex constitutive equations have been developed. With the degree of complexity the number of material constants rises that have to be defined for every specific soil.[8] Constitutive laws are e.g. HOOKE's law (linear elastic, isotropic), CamClay (elastic-plastic), Mohr-Coulomb (elastic-plastic) and hypoplasticity.

Hypoplasticity

Hypoplasticity[9] expresses stress increments as a function of given strain increments and of the actual state (stress, void ratio). The formulation is non-linearly in the stretching tensor $\mathbf{D}$ (symmetric part of the velocity gradient). The equation is homogeneous in the CAUCHY stress tensor $\mathbf{T}$.

There are various versions of hypoplasticity, see e.g. review by Fellin and Kolymbas[10] or by Wu 2000.[11] Several improvements of the hypoplastic constitutive law were developed since the first version 1977 by Kolymbas. A major advancement was published 1990 by Wu and Kolymbas, which stands now as the basis for further developments of hypoplasticity.

The advantages of hypoplasticity are that void ratio and stress are in general the only state variables of the soil, and the material parameters can be easily obtained from laboratory tests. Hypoplasticity can be used for the simulation of many aspects of soil mechanical behaviour, such as stress and density dependent soil behaviour, and non-linear stress-strain behaviour.

---

[8]Kolymbas, D.: Geotechnik – Bodenmechanik und Grundbau, Springer-Verlag 1998, p. 199 et sqq.

[9]A comprehensive summary about hypoplasticity was published by Kolymbas 2000: Kolymbas, D.: Introduction to Hypoplasticity, in *Advances in Geotechnical Engineering and Tunnelling*, Vol. 1, A.A. Balkema, Rotterdam, 2000

[10]Fellin, W., Kolymbas, D.: Hypoplastizität für leicht Fortgeschrittene (Hypoplasticity for the lower intermediate), *Bautechnik*, 12 (**79**) 2000, pp. 830-841

[11]A concise report about the history can be found in:
Wu, W., Kolymbas, D.: Hypoplasticity then and now, in *Constitutive Modelling of Granular Materials*, Springer-Verlag Berin Heidelberg, 2000, pp. 57-105,
which is, therefore, here only short summarized

Calibration (determination of material parameters):

For the hypoplastic constitutive law – version v. Wolffersdorf – eight material parameters have to be defined[12]:

- critical friction angle φ_c
- granulate hardness h_s and exponent n, cf. equ. (2.4)
- limit void ratios at zero stress e_{d0} (densest), e_{c0} (critical) and e_{i0} (loosest)
- exponents α (characterises influence of density on the peak friction angle) and β (for scalar multiplier of the stress rate, so that stress rate increases with increasing density)

The critical friction angle φ_c can be obtained from shear or triaxial tests, but it is hardly feasible due to inevitable inhomogeneous deformations of the specimen. The critical state is reached after a big shear strain. A simple method for the determination of φ_c for cohesionless materials is to pour a cone from a funnel; the resulting slope is approximately to the critical friction angle φ_c. Limits of applicability are given by Herle[12].

The granulate hardness h_s is a density-independent reference pressure; it is the only dimensional variable in hypoplasticity. Its magnitude is mainly affected by grain fabric, grain shape, grain size distribution and the grain size. h_s is closely connected to the exponent n. Both can be derived via numerical regression from compression test results, such as isotropic or oedometric compression, using equation[13]

$$\frac{e_i}{e_{i0}} = \frac{e_c}{e_{c0}} = \frac{e_d}{e_{d0}} = \exp\left[-\left(\frac{3p_s}{h_s}\right)^n\right] \quad , \tag{2.4}$$

with p_s being the mean effective stress. Various combinations of h_s and n can lead to a good fit, therefore bounds of n were proposed by Herle[14],

$$0 \le n \le 0.66 \quad . \tag{2.5}$$

---

[12]Taken over from the dissertation of Herle (1997):
Herle, I.: Hypoplastizität und Granulometrie einfacher Korngerüste (Hypoplasticity and granulometry of simple granular structures), PhD Dissertation, University of Karlsruhe, 1997

[13]Gudehus, G.: A comprehensive constitutive equation for granular material, *Soils & Foundations*, 1 (**36**) 1996, pp. 1-12

[14]Herle, I.: A relation between parameters of a hypoplastic constitutive model and grain properties, in Proc. of the Fourth International Workshop on Localization and Bifurcation Theory for Soils and Rocks, Gifu, Japan, 28. Sept.-2. Oct. 1997, A.A. Balkema, Rotterdam, 1998

The considered stress range in the compression test should be sufficiently wide to assure a combination of h_s and n which leads to realistic reproduction of the compression curves using equ. (2.4). n can be approximated from an empirical relation proposed by Herle[12]:

$$n = f\left(U/(d_{50}/d_0)^{1/3}\right) \quad , \tag{2.6}$$

with $d_0 = 1$ mm being a reference grain size. For Ottendorf-Okrilla sand this approximation leads to $n \approx 0.21$, read from a empirical diagram of Herle[12], which was derived from various experimental results.

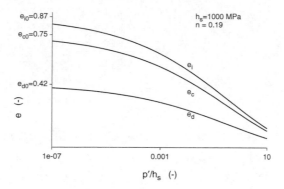

Fig. 2.13: Pressure dependent void ratio

The limit values of void ratios at zero stress (Fig. 2.13) can be estimated from laboratory test of the minimum and maximum void ratio:

- The upper bound of possible void ratios is e_i. e_{i0} can be estimated from the maximum void ratio e_{max} (loose packing of soil): $e_{i0} \approx 1.15 e_{max}$ (Herle[12]).
- The minimum void ratio e_{min} from standard densification test (vibrating with a fork, see section 3.3.4) gives approximately e_{d0}. The value acquired from shaking table test seems to be more convenient, as it leads to higher compaction.

Critical void ratio e_{c0} at zero pressure is usually assumed as e_{max}, and can be extrapolated using equ. (2.4), whereas the aforementioned difficulties of critical state must be pointed out.

The exponent α can be derived considering the peak state from

$$\alpha = \frac{\ln\left(6\,\frac{(2+K_p)^2 + a^2 K_p(K_p - 1 - \tan\nu_p)}{a(2+K_p)(5K_p - 2)\sqrt{4 + 2(1 + \tan\nu_p)^2}}\right)}{\ln(r_e)} \quad , \tag{2.7}$$

with the peak values of the earth-pressure coefficient K_p, the dilatancy angle ν, the friction angle φ_p and the pressure dependent relative density r_e

$$K_p = \frac{T_1}{T_2} = \frac{1 + \sin\varphi_p}{1 - \sin\varphi_p} \quad , \tag{2.8}$$

$$\tan\nu_p = -\frac{D_1 + 2D_2}{D1} \quad , \tag{2.9}$$

$$\varphi_p = \sin^{-1}\left(\frac{T_1 - T_2}{T_1 + T_2}\right)_p \quad , \tag{2.10}$$

$$r_e = \frac{e - e_d}{e_c - e_d} \quad . \tag{2.11}$$

The exponent α is typically in the range 0.1 to 0.3. The experimental determined peak friction angle also depends on the test type (e.g. triaxial, biaxial). Herle[12] suggests triaxial tests on initially dense specimens for a precise determination of α.

The exponent β can be derived, e.g. from an isotropic consolidation of an initially loose specimen using

$$\beta = \frac{\ln\left[\frac{E_s}{f_b(3 + a^2 - f_d a\sqrt{3})}\right]}{\ln(e_c/e)} \quad , \tag{2.12}$$

with the pressure- and density-dependent stiffness modulus $E_s = \dot{T}_1/D_1$ and

$$a = \frac{\sqrt{3}(3 - \sin\varphi_c)}{2\sqrt{2}\sin\varphi_c} \quad . \tag{2.13}$$

The isotropic compression is the theoretically simplest case, because all pressure rates and strain rates are identical and the hypoplastic constitutive law reduces to $\dot{T}_1 = f_b f_e \left(3 + a^2 - f_d a\sqrt{3}\right) D_1$. For many granular soils the exponent β can be assumed as 1.

Hypoplastic parameters for Ottendorf-Okrilla sand and Soiltron

The critical friction angle was estimated from the triaxial tests (see section 4.1) as $\varphi_c \approx 32\ldots33°$. From a series of tests for the estimation of the average angle of repose[15] φ_c was derived as 32.6°.

Test no.	measured angle of repose (in °) at point				average
	90°	**180°**	**270°**	**360°**	value
1	31.9	34.5	32.4	32.6	32.9
2	33.4	33.6	32.6	32.0	32.9
3	31.3	33.3	32.3	34.2	32.8
4	32.1	33.2	34.2	31.7	32.8
5	31.7	33.7	30.1	33.0	32.1
6	31.3	33.4	33.1	31.7	32.4
					32.6

Tab. 2.2: Angles of repose (Ottendorf-Okrilla sand)

The limit void ratios are measured in standard laboratory tests, see section 3.3.4, as $e_{max} = 0.75$, $e_{min} = 0.42$.

$$e_{i0} \approx 1.15 e_{max} = 0.86$$
$$e_{c0} \approx e_{max} = 0.75$$
$$e_{d0} \approx e_{min} = 0.42$$

h_s and n were estimated from oedometric compression tests of dry sand, see Fig. 2.14. The specimens were prepared with the help of 99% methyl alcohol, as described in section 5.1. With the help of this technique very loose specimens — even looser than e_{min} — could be prepared. The alcohol evaporated before the tests, therefore, no capillary forces interfered with the measurements.

---

[15]The author is indebted to M. Mähr of the Institute of Geotechnical and Tunnel Engineering, University of Innsbruck to make the data available.

n can be acquired using the compression index[16] $C_c = \mathrm{d}e/\mathrm{d}\ln(\sigma/\sigma_0)$

$$n = \frac{\ln\left(\dfrac{e_{p1}C_{c2}}{e_{p2}C_{c1}}\right)}{\ln\left(\dfrac{p_{s2}}{p_{ps1}}\right)} \quad , \tag{2.14}$$

with e_{pi} being void ratios and C_{ci} being the compression index (slope of compression line) at the pressures p_{si}, see table 2.3. A rough estimation using regression of experimental data on several sands was given by Herle[16]. The diagram of Herle leads to $n \approx 0.21$ for Ottendorf-Okrilla sand and Soiltron. In this study n was calculated using equ. (2.14).

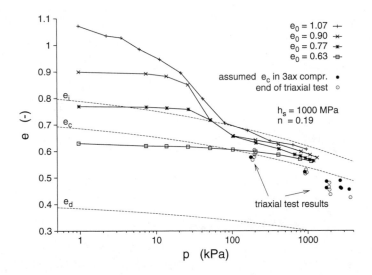

Fig. 2.14: Oedometric compression of Ottendorf-Okrilla sand and final value of void ratio in triaxial test, where • marks the void ratio at the assumed critical state (see Fig. 4.1) and ∘ the void ratio at the end of the triaxial test

The sensitivity of the exponent n is high. Therefore a wide range of pressure was chosen for the estimation of this value. The granulate hardness was estimated using Herle's fomula

$$h_s = 3p_s \left(\frac{n\,e_p}{C_c}\right)^{1/n} \quad . \tag{2.15}$$

[16]Herle, I.: Hypoplastizität und Granulometrie einfacher Korngerüste (Hypoplasticity and granulometry of simple granular structures), PhD Dissertation, University of Karlsruhe, 1997

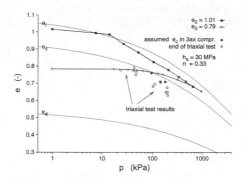 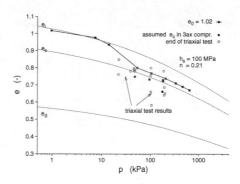

Fig. 2.15: Oedometric compression of Soil-tron 1 (additive PS), description as in Fig. 2.14

Fig. 2.16: Oedometric compression of Soil-tron 2 (additive perlite), description as in Fig. 2.14

Material	e_0	T_{1i}	e_{pi}	C_{ci}	n	h_s
	(-)	(kPa)	(-)	(kPa)	(-)	(MPa)
Ottendorf-Okrilla sand	0.77	160	0.656	0.027	0.19	1000
		1840	0.563	0.037		
Soiltron 1	1.01	82	0.88	0.048	0.20	100
		1002	0.68	0.091		
Soiltron 2	1.02	322	0.74	0.053	0.21	100
		1002	0.67	0.061		

Tab. 2.3: Readings from the compression curves and calculated C_{ci}, n and h_s (data of representative tests, i.e. tests where $e_0 \approx e_{c0}$)

As can be seen from table 2.3, the granulate hardness can be drastically decreased with the intermixture of soft particles. This can be attibuted to the interruption of the force chains.

The exponent α was estimated using equ. (2.7). The parameters for this equation can be found in table 2.4. As can be seen the values are quite same. The dilatancy angle $\tan \nu_p$ was calculated using

$$\tan \nu_p = 2 \frac{K_p - 4 + 5AK_p^2 - 2AK_p}{(5K_p - 2)(1 + 2A)} - 1 \quad , \tag{2.16}$$

with

$$A = \frac{a^2}{(2 + K_p)^2} \left[1 - \frac{K_p(4 - K_p)}{5K_p - 2} \right] \quad . \tag{2.17}$$

Material	cell pressure σ'_r (kPa)	friction angle $\bar{\varphi}_p$ (°)	dilatancy angle $\bar{\nu}_p$ (-)	relative density $\bar{r}_e$ (-)	α (-)
O.-O. sand	100	39.8	21.6	0.50	0.19
Soiltron 1	100	36.7	13.6	0.65	0.20
Soiltron 2	100	37.1	14.6	0.65	0.21

Tab. 2.4: Estimation of hypoplastic parameter α (average values)

The exponent β can be calculated using equ. (2.12) from the isotropic compression of a initially dense specimen. The determination of β was performed iteratively, because f_b already contains β

$$f_b = \frac{h_s}{n} \left(\frac{e_{i0}}{e_{c0}}\right)^\beta \frac{1 + e_i}{e_i} \left(\frac{3 p_s}{h_s}\right)^{1-n} \left[3 + a^2 - a\sqrt{3}\left(\frac{e_{i0} - e_{d0}}{e_{c0} - e_{d0}}\right)^\alpha\right]^{-1}. \quad (2.18)$$

f_d is defined as

$$f_d = r_e^\alpha = \left(\frac{e - e_d}{e_c - e_d}\right)^\alpha . \quad (2.19)$$

The triaxial test SAP1g had an initial void ratio $e_0 = 0.522$ ($r_e = 0.30$) and was isotropically compressed to $p_s = 100$ kPa, see Fig. 2.17. With $E_s = 64.5$ MPa the exponent was calculated as $\beta = 2.25$.

(Soiltron 1, test SO1P1g: $e_0 = 0.721$, $r_e = 0.40$, isotropic compressed to $p_s = 100$ kPa, $E_s = 30.5$ MPa $\rightsquigarrow \beta = 3.48$; Soiltron 2, test SO2P1e: $e_0 = 0.696$, $r_e = 0.14$, isotropic compressed to $p_s = 100$ kPa, $E_s = 26.5$ MPa $\rightsquigarrow \beta = 2.25$)

The test results of the isotropic compressions of Soiltron 1 and 2 are shown in Chapter 4.

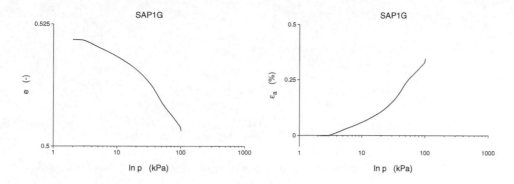

Fig. 2.17: Isotropic compression of Ottendorf-Okrilla sand

Numerical simulations with the obtained hypoplastic parameters

The hypoplastic parameters were checked using the single element program *TRIAX* by D. Mašin[17] with calculation of triaxial tests and comparison with results of some triaxial tests, see Fig. 2.18. Details of the triaxial tests can be found in Chapter 4.

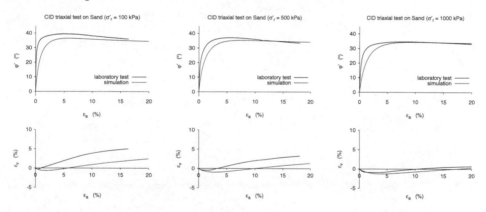

Fig. 2.18: Numerical simulations of triaxial tests with the obtained hypoplastic parameters, triaxial tests at 100, 500 and 1000 kPa confining pressure

From the comparisons in Figures 2.18-2.20 can be seen that the simulations give a reasonable fit of the stress-strain and volumetric behaviour of Ottendorf-Okrilla sand in triaxial compression at various confining pressures. The predictions for higher pressures are closer to the experimentally observed results. The reliability of the triaxial test results is proven in Chapter 3. Note that h_s and n were obtained for only few oedoemtric compression test results. Other combinations of these parameters are possible.

---

[17]http://www.natur.cuni.cz/~masin/triax/triax.html

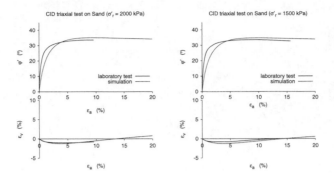

Fig. 2.18: (continued) Numerical simulations of triaxial tests with the obtained hypoplastic parameters, triaxial tests at 1500 and 2000 kPa confining pressure

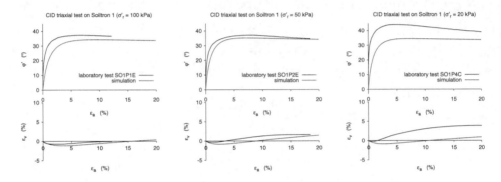

Fig. 2.19: Comparison of laboratory results and numerical simulations of triaxial tests with Ottendorf-Okrilla sand and Soiltron 1

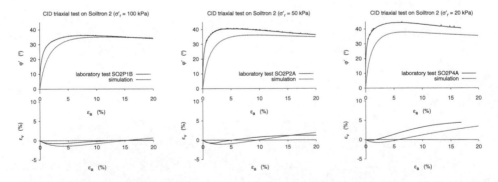

Fig. 2.20: Comparison of laboratory results and numerical simulations of triaxial tests with Ottendorf-Okrilla sand and Soiltron 2

The hypoplastic parameters for Ottendorf-Okrilla sand and Soiltron are summarised in table 2.5. Note, that the parameters are reasonable for the observed pressure range, which was proved with comparison of test results and simulations. Since many combinations of h_s and n can give good agreement with experimental compression curves and due to the fact that these two parameters subsequently influence α and β, if calculated after equations (2.7)-(2.12), other sets of parameters might be found.

| **Material** | φ_c | h_s | n | e_{d0} | e_{c0} | e_{i0} | α | β |
	[°]	[MPa]	[-]	[-]	[-]	[-]	[-]	[-]
O.-O. sand	32	1000	0.19	0.42	0.75	0.86	0.19	2.25
Soiltron 1	32	100	0.20	0.54	1.00	1.15	0.20	3.48
Soiltron 2	32	100	0.21	0.63	1.00	1.15	0.21	2.25

Tab. 2.5: Hypoplastic parameters of Ottendorf-Okrilla sand and Soiltron

For the validation of area correction assumptions taken for the evaluation of triaxial tests (section 3.2.1.3), a numerical study on the influence of end restraints on the deformation of the triaxial specimen was conducted using hypoplasticity (section 3.3.5).

Chapter 3

Triaxial tests — devices, processes and limitations

3.1 Triaxial test device

For the systematic investigation of sand and Soiltron, isotropically-consolidated drained (CID) triaxial tests with dry specimens have been carried out in the Soil Mechanics Laboratory at the Charles University in Prague.

The tests were conducted in a computer controlled, standard triaxial apparatus (WFi TriTech 50 kN) with internal load cell (capacities up to 25 kN) on 75 mm diameter specimens with lubricated end platens and an aspect ratio (height/diameter) of approximately 1.

To avoid piston friction, force was measured inside the triaxial cell. The cell pressure was controlled with a GDS standard pressure/volume controller.

3.1.1 End conditions

The specimens had an 1:1 aspect ratio and were bedded on lubricating layers. The lubrication consisted, as proposed by Rowe[1], of two 0.3 mm thin rubber membranes and standard silicone grease (Spanjaard[2]) aplied on the polished platens. In the middle of the lower platen was a central drainage hole with a small porous stone. The filter stone in the lower platen is necessary (i) for drainage and (ii) for lateral fixation of the specimens. Therefore, this hole was not covered by the lubrication film. Layers adjacent to the specimen were equipped with radial cuts to minimize the restraint of the specimen's lateral deformation.

---

[1]Rowe, P.W., Barden, L.: Importance of free ends in triaxial testing, *Journal of Soil Mechanics and Foundation Division*, ASCE, 1 (**90**) 1964, pp. 1-27

[2]Spanjaard Ltd., Hill House, 1 Little New Street, London EC4A 3TR

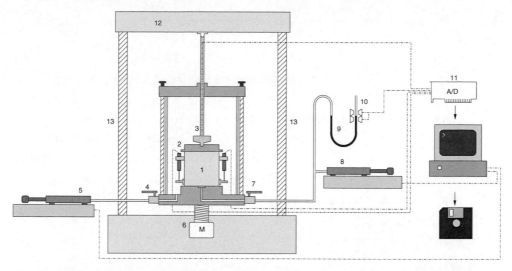

Fig. 3.1: General configuration of the test equipment: 1 specimen, 2 local displacement transducer (LVDT), 3 load cell and loading ram, 4 stop valve for cell liquid, 5 GDS standard pressure/volume controller for cell liquid, 6 step motor drive, 7 stop valve for pore-air (drainage), 8 GDS standard pressure/volume controller for pore-air, 9 U-shaped observation tube, 10 photoelectric sensors, 11 A/D-card for data acquisition, 12 triaxial load frame (cross head), 13 tie rods

These provisions were taken, to improve the homogeneous stress distribution within the specimens. However, free ends have also disadvantages in addition to the difficult setup of the test. With increasing axial stress, the grease is squeezed out. This causes bedding error and, probably the effectiveness of the lubrication is reduced.

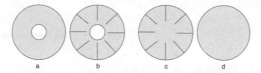

Fig. 3.2: Latex rubber discs for the lubrication of the end platens; a) and b) for the bottom plate, c) and d) for the top plate; the membranes adjacent to the soil specimen (b) and (c) are incised to ease the radial deformation of the specimen

Preparation of the lubricated platens: First the platens were thoroughly cleaned and the first layers of grease were uniformly applied. The membranes (Fig. 3.2a and d) were laid onto the grease and pressed so that no visible air bubble was entrapped under it. The second grease layers were spread uniformly on the membranes and the second layers of incised membranes (Fig. 3.2b and c) were placed onto them.

The magnitude of bedding error due to lubrication can be assessed according to several methods:

- El-Sohby, cited in Sarsby[3], estimated the bedding error by placing an additional membrane layer of same size and composition in the midheight of the specimen and compared the axial deformation of this specimen with those, prepared without this additional lubrication layer. The difference of these two measurements was an estimation of the bedding error.
- McDermott, also cited in Sarsby, also used additional layers within the specimen, but two sets at midheight or 1/3-points, respectively, to evaluate the bedding error.

In this study a rigid dummy instead of the soil specimen was used to determine the bedding error. However, as can be seen in section 3.2.1.6, this method gives no reliable results, due to the different surfaces of dummy and soil specimen and the resulting different penetration into the lubricating layers.

Furthermore, inhomogeneous deformation was observed even with the use of lubricated end platens. Often, barreling or formation of elephant foot took place. The evaluation of these tests is described in section 3.2.

3.1.2   Axial deformation measurement

The axial deformation of the specimens was measured globally by a dial gauge on top of the cell, and locally using two diametrically mounted local displacement transducers LVDT[4] attached directly to the membrane, see Fig. 3.1.

For the verification of the area corrections (see section 3.2.1.3) axial deformations were measured remotely and evaluated using the PIV method, see section 3.1.4.

3.1.2.1   'Global' measurement

The axial deformation can be assessed using a dial gauge placed on top of the triaxial cell. Due to bedding error and compliance of the loading system, this

---

[3]Sarsby, R.W., Kalteziotis, N., Haddad, E.H.: Compression of "free ends" in triaxial testing, *Journal of the Geotechnical Engineering Division*, 1 (**108**) 1982, pp. 83-107

[4]LVDT type RDP D5/200WRA/131, RDP Electronics Ltd

measurement is not reliable when working with lubricated ends. Therefore, it was decided to consider the 'global' axial deformation measurement only at a later stage of the test, when the two local deformation measurements began to deviate from eachother, see section 3.2.1.3.

A calibrated digital dial gauge Mahr[5], Fig. 3.3, with a range of 25 mm (max $\varepsilon_a \approx 33\%$) and a resolution of 0.001 mm was used for the 'global' defomation measurement. The calibration sheet of the manufacturer certified a maximal deviation from linearity of 2 μm.

Fig. 3.3: Mahr dial gauge for the 'global' deformation measurement

3.1.2.2 Local measurement

Two LVDTs, manufactured by RDP Electronics Ltd[6], were mounted diametrically to measure axial deformation. They were mounted with light-weight brackets glued with cyanoacrylate onto the membrane. They consist of a core and a coil, the linear measuring range[7] was approximately 10 mm. The gauge length between the mounts was typically 50 mm, which leads to a maximum axial strain range of max $\varepsilon_a \approx 20\%$. The cores are resting on the lower mount and allow the specimen to barrel without restraint. The two seperate

---

[5]Mahr GmbH, Esslingen, Germany

[6]Grove Street, Heath Town, Wolverhampton UK, Telephone: +44-1902-457512, Fax: +44-1902-452000

[7]With the LVDTs a calibration sheet was delivered. The linearity was certified with an accuracy of $\leq$0.10%

measurements of the LVDTs were averaged. They were used as long as the deformation was homogeneous, see section 3.2.1.3.

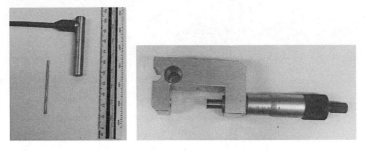

Fig. 3.4: LVDT (RDP Electronics Ltd.); Micrometer for calibration

3.1.3   Volume change measurement

For the measurement of volume change of dry specimens a similar device as the one by Bishop & Henkel[8] was used (Fig. 3.5a). Bishop & Henkel proposed to observe an oil level in an U-shaped tube, connected to the drainage line. When keeping the oil level constant, a direct measurement of volume change is possible with the burette.

Without using a burette this measurement is possible, using a device for keeping the air pressure within the specimen equal to, say, atmospheric pressure. For this reason an U-shaped observation tube was constructed, which can be connected to the drainage and a GDS controller, Fig. 3.5b.

The U-shaped observation tube was filled with coloured ethanol. The liquid level was observed using two photoelectric sensors mounted to the side of the tube connected to the atmosphere. The other side of the tube is connected to the specimen drainage and with a branch connection to a GDS standard pressure/volume controller (resolution: ± 1 mm^3). The air expelled from the specimen due to volume changes moves the menisk in the U-tube. This is reversed by appropriate movement of the piston of the GDS controller. This procedure makes possible to measure the volume changes in a way that is not biased by the compressibility of air.

The pressure within the specimen was initially atmospheric. Consolidation was conducted using the 'ramp cycle' function of the GDS controller (Fig. 3.1,

---

[8]Bishop, A.W. and Henkel, D.J.: The Measurement of Soil Properties in the Triaxial Test, Second Edition, Edward Arnold (Publishers) Ltd., London, 1962

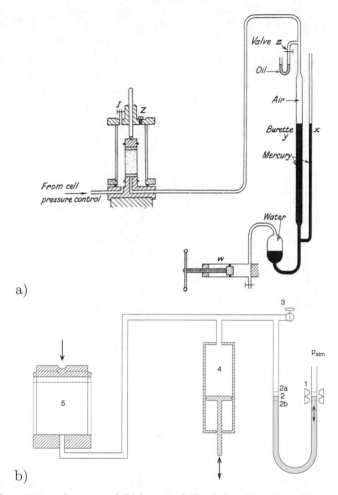

a)

b)

Fig. 3.5: a) Experimental setup of Bishop and Henkel, taken from Bishop & Henkel; b) Schematic sketch of the used experimental setup to control the pore-air pressure: 1 photoelectric sensors, 2 mean water level in observation tube, 2a and 2b highest and lowest menisk levels, 3 stop valve, 4 GDS standard pressure/volume controller, 5 triaxial specimen

component 5) in steps of 2 sec/kPa, to avoid pore-air excess pressures. Pore-air excess pressure can be recognized at the U-tube, as the liquid level would then exceed the upper photoelectric sensor (due to compression of the specimen, air is expelled from the specimen, hence the liquid level rises in the U-tube at the side open to atmosphere) and would lead to a misinterpretation of the measured volume change.

The distance between the two photoelectric sensors was 6 mm, the tube inner diameter was 2.6 mm. The precision of this method is controlled by the accuracy of the U-tube readings, which is ± 15 mm^3 (corresponding to $\Delta\varepsilon_v \pm 0.01\%$,

Test	cell pressure σ_r (kPa)	initial void ratio e_0 (-)	test condition	max. temp. change ΔT (K)
D100	100	0.526		0.1
D500	500	0.513		0.1
D1000	1000	0.469	dry	0.4
D1500	1500	0.496		0.7
D2000	2000	0.494		0.3
S100	100	0.469		0.4
S500	500	0.481	water	0.2
S1000	1000	0.468	saturated	0.2
S1500	1000	0.444		0.5
S2000	1000	0.497		0.1

Tab. 3.1: Overview of tests to assess the performance of the volume change measurement

assuming a specimen volume of 330 cm^3 and an average void ratio of e=0.5).

Since air is very sensitive to temperature changes, isothermal conditions were aimed. The laboratory is located in the basement of the building. One test lasts about 2-4 hours, including consolidation and shearing stages of the test. The maximum deviation within this time was 0.7°C. It was assumed that this temperature change was neglible.

For every cell pressure applied, one dry and one water-saturated test was conducted, the results of which are shown in Fig. 3.6. As can be seen from the stress-strain curves and volumetric strain vs. axial strain, a good coincidence of dry and water saturated tests was achieved. The test results demonstrate that the proposed method is of comparable precision to the standard method used for water saturated specimens.

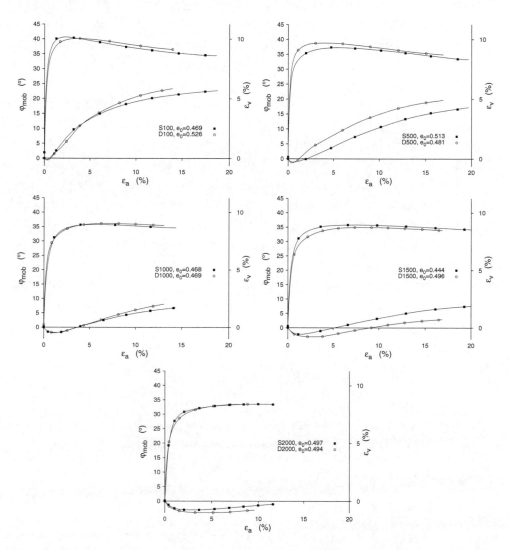

Fig. 3.6: Normalized stress-strain and volumetric strain curves; S marks the water-saturated tests, D the dry tests, the number refers to the cell pressure

3.1.4 Remote measurement of axial and volumetric deformations using PIV

Particle Image Velocimetry (PIV) can be used to track the movement of particles. It was originally developed for the evaluation of fluid flows, mixtures of solid-liquid, solid-gas, liquid-gas and spray/droplet laden flows. Particles are seeded into the flow and illuminated with laser pulses in a thin sheet, see e.g. Grant.[9] Flow fields (velocity vectors) are constructed from consecutive images (analog or digital) by pointwise interrogation of small parts of the obtained images. The application of PIV techniques to solid mechanics is relatively new. For example Nübel[10], Nübel & Weitbrecht[11] and White *et al.*[12] did extensive investigations in that field recently. The accuracy depends on many factors, such as density of particles, flow velocity, velocity range (velocity distribution within the region of interest), sub-pixel peak location or so-called out-of-pattern effects, see e.g. Adrian[13], Huang *et al.*[14], Moreno *et al.*[15], Olsen *et al.*[16], Udrea *et al.*[17] Using the image shifting method, errors arising from low correlation (out-of-pattern effects), due to particle displacements, can be reduced. Methods such as centre-of-mass, parabolic curve-fitting and Gaussian curve-fitting are reported to give good results for correlation with sub-pixel accuracy.[14]

---

[9]Grant, I., Smith, G.H.: Modern developments in Particle Image Velocimetry, *Optics & Lasers in Engineering*, 3-4 (**9**) 1988, pp. 245-264

[10]Nübel, K.: Experimental and numerical investigation of shear localization in granular material, PhD University of Karlsruhe, 2002, Vol. 159

[11]Nübel, K., Weitbrecht, V.: Visualization of localization in grain skeletons with Particle Image Velocimetry, *Journal of Testing and Evaluation*, 4 (**30**) 2002, pp. 322-328

[12]White, D.J., Take, W.A, Bolton, M.D., Munachen, S.E.: A deformation measuring system for geotechnical testing based on digital imaging, close-range photogrammetry, and PIV image analysis, Proceedings of the 15th International Conference on Soil Mechanics and Geotechnical Engineering, Istanbul, Turkey, 2001, Vol. 1, pp. 539-542

[13]Adrian, R.J.: Dynamic ranges of velocity and spatial resolution of particle image velocimetry, *Measurement Science and Technology*, 12 (**8**) 1997, pp. 1393-1398

[14]Huang, H., Dabiri, D., Gharib, M.: On errors of digital particle image velocimetry, *Measurement Science and Technology*, 12 (**8**) 1997, pp. 1427-1440

[15]Moreno, D., Mendoza Santoyo, F., Funes-Gallanzi, M., Fernandez Orozco, S.: An optimum data display method, *Optics & Laser Technology*, 2 (**32**) 2000, pp. 121-128

[16]Olsen, M.G., Adrian, R.J.: Brownian motion and correlation in particle image velocimetry, *Optics & Laser Technology*, 7-8 (**32**) 2000, pp. 621-627

[17]Udrea, D.D., Bryanston-Cross, P.J., Lee, W.K., Funes-Gallanzi, M.: Two sub-pixel processing algorithms for high accuracy particle centre estimation in low seeding density particle image velocimetry, *Optics & Laser Technology*, 5 (**28**) 1996, pp. 389-386

The aim of this study was the investigation of slow motions of grains during a triaxial test and to measure vertical and radial displacements of the specimens to prove the area assumptions (section 3.2.1.3) for the evaluation of the triaxial tests (section 4.1). Here, grains of a soil specimen in a triaxial test are visible through the more or less transparent enclosing membrane and the water-filled triaxial cell. Deformations resulting from loading of the specimen can be evaluated from the interrogation of sub-windows of the whole image and compare the amount of match between two consecutive images using cross-correlation. The application of PIV to triaxial tests was applied by Rinawi[18] in his diploma thesis.

The program MatPIV v.1.6, available under GNU public license from Johan Kristian Sveen[19,20], was used. This program was developed for evaluation of fluid flow and was adapted within several diploma theses[18,21] to soil and triaxial tests. Images were digitally recorded using a CCD (charge coupled device) camera. The calculations are done using subroutines of the mathematical programming environment MatLAB v.13.[22] For the evaluation of triaxial tests with the PIV method, the program *SoilPIV*, subdivided into two program modules for MatLAB, was developed by Rinawi[18].

The errors due to aliasing, bias errors and outliers must be removed. Aliasing: windows are assumed as periodic, therefore the correlation peak could appear on the opposite side. This can be removed by increasing the subwindow size. Bias errors: if the displacement is large, the resulting correlation is low. Bias effects are removed using weighting functions. Outliers (spurious vectors) are easy to detect due to their small correlation peak and their random distribution.

PIV works with pixel coordinates, displacements can be calculated transforming pixel coordinates to world coordinates with special subroutines. Several filters (signal-to-noise, global and local filters) are applied to the images to remove spurious vectors. A comprehensive description of the filters can be found in Sveen[20] and in Rinawi.[18]

---

[18]Rinawi, W.: Particle Image Velocimetry (PIV) applied on triaxial tests, MSc. thesis, Institute of Geotechnical and Tunnel Engineering, University of Innsbruck, 2004

[19]available at http://www.math.uio.no/~jks/matpiv

[20]Sveen, J.K.: An introduction to MatPIV v. 1.4, http://www.math.uio.no/~jks/

[21]Fromm, H.: Experimentelle Überprüfung von Oberflächensetzungen infolge Ringspaltes und seiner Verpressung, MSc. thesis, Institute of Geotechnical and Tunnel Engineering, University of Innsbruck, 2002

[22]MathWorks Inc. http://www.mathworks.com

For the evaluation of deformations in triaxial tests using PIV, the following conditions and certain restrictions must be taken into account:

– The illumination must be free of reflexions, therefore reflecting parts at the equipment had to be avoided. E.g. the camera fixing and lamp tripods were made of dark material. Windows were dimmed out or tests were conducted during night, so that the illumination during the tests was constant. An angular rotation of approximately 30° of the lights to the optical axis of the camea proved to be best for a uniform illumination of the specimens.

– To avoid heating of the cell water, and subsequent change of refraction, the lamps were only switched on when taking pictures.

– The higher the effective cell pressure was chosen, the better the grains were visible due to membrane penetration. For further informations concerning membrane penetration see section 3.2.1.7.

– PIV requires for a higher accuracy a certain 'velocity' of the investigated displacements. In the triaxial test the velocity is relatively small and the grains are close together. In the standard application of PIV for flows, the particles are seeded in certain distances, therefore they are more easy to track.

– Since here only one digital camera was used for a 2D PIV-analysis, the mathematical model to calculate 3D deformations of the specimen some assumptions were taken: The deformation of the specimens was assumed as radially for the discretized surface (Fig. 3.12) and it was assumed that the optical axis and the center of the specimen coincide. With the first attempt to investigate the specimen deformation in triaxial tests with PIV, the calculated (PIV-modelled) specimen is composed of a user-defined number of frustum sectors.

– Images have to be saved in single matrices to perform the correlation calculations in MatLAB, therefore it is needed to store the pictures in a pixelwise indexed image format, like TIFF (tagged image file format). This was possible using a Minolta Dimage 7i digital camera. This leads with 5 Megapixel resolution (2560×1920 px) to approximately 14.1 MB size for each image in, so-called, super fine quality.

– Image compression was not necessary due to the relatively slow velocity of the triaxial tests, and therefore time was sufficient to store images on the memory chip of the camera. Image compression to e.g. JPEG changes informations stored in the image. With increasing image compression, spurious velocity vectors are more often encountered. JPEG-compression is a technique, that decreases image quality with increasing image compression rate. There are also lossless image compression methods, but they were not available and needed at that camera, able to save one picture in highest quality within approximately 30 seconds (corresponds here to ca 0.15% axial strain). The effects of compression on PIV-results are summarized in Cenedese *et al.*[23]

— Colored marks on the membrane can be used to increase the texture of the surface and, therefore, the accuracy, see e.g. White *et al.*[24] This is useful, e.g., for fine soils or clays. This was not necessary in this case, because the grains of sand have an inherent texture, as can be seen in Fig. 3.7, where the histogram of the texture shows a large bandwidth, is presented.

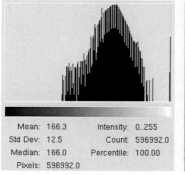

Mean: 166.3   Intensity: 0..255
Std Dev: 12.5   Count: 596992.0
Median: 166.0   Percentile: 100.00
Pixels: 596992.0

Fig. 3.7: Histogram of pixel intensity of a patch of sand in the triaxial cell

3.1.4.1 Experimental setup

For the application of PIV the triaxial testing equipment had to be adapted. One of the cell screw connections had to be removed, to be able to see the whole specimen. For safety reasons the cell pressure was limited to 1 MPa. The digital camera was fixed in a distance of 750 mm to the cell. The autofocus function of the digital camera was disabled, as different apertures and

[23] Cenedese, A., Pocecco, A. Querzoli, G: Effects of image compression on PIV and PTV analysis, *Optics & Laser Technology*, 2 (**31**) 1999, pp. 141-149

[24] White, D.J., Take, W.A., Bolton, M.D.: Measuring soil deformation in geotechnical models using digital images and PIV analysis, 10th International Conference on Computer Methods and Advances in Geomechanics, Tucson, Arizona 2001, Balkema, pp. 997-1002

focal lengths in the pictures would lead to useless images. Therefore, lighting conditions have to be constant during the tests. The illumination consisted of two 1000 Watt lamps, fixed on tripods shifted 30° towards the optical axis of the digital camera in a height of approximately 1 m above the camera. The camera was aligned perpendicular to the specimens midheight.

Due to the cylindrical dimension of the specimens and the 2D characteristic of the digital images, all measurements had to be referred to reference points, arbitrarily defined in the center of the pedestal. These reference points (equispaced crosshairs) were located on a plate in the center of the triaxial cell, at the same position, where the specimen is built later (Fig. 3.8). This plate was photographed before every test, to define the world coordinates, which enable the conversion from pixel coordinates to 'real' positions.

Fig. 3.8: Reference points on a centered vertical plate

The digital camera was automatically controlled to take pictures by the data acquisition and control program for the triaxial tests. The pictures were taken in manually set time steps. Hence, locally and globally measured deformations can be proper associated to the images for comparison reasons of the different deformation measurements.

3.1.4.2   Evaluation of tests

To apply MatPIV to the evaluation of triaxial tests, some modifications and additional calculations were required, due to the two-dimensional nature of standard PIV applications. The cell and specimens are cylindrical, water and perspex refract the light, therefore the visible deformations are not the real ones. The apparent dimensions, see Fig. 3.9, of the specimen depend on the refraction indices of the water-perspex and perspex-air interfaces.

Fig. 3.9: Triaxial specimen: without cell and with perspex cell and cell water

Fig. 3.10: Refraction of light

Therefore, the routines 'virt2real' and 'real2virt' were developed by Rinawi[18], to convert photographic dimensions into the spatial location of discrete points on the surface of the specimen.[25] These routines are sub-programs used in the MatLAB programming environment and base on trigonometric calculations and SNELL's law of refraction

$$\frac{\sin \alpha_1}{\sin \alpha_2} = \frac{n_2}{n_1} \quad , \tag{3.1}$$

with the angles α_i according to Fig. 3.10 and the refraction indices n_i. Typical values are given in table 3.2.

Medium	Refraction index n
Air	$n_a = 1.0003$
Water	$n_w = 1.33$
Perspex (triaxial cell)	$n_p = 1.49$

Tab. 3.2: Typical refraction indices after Rinawi[18]

Module I (Pre-processor): The main purpose of this module is to process the acquired images with MatPIV. Further purposes are:

---

[25] The interested reader is refered to this work, which is available by the Institute of Geotechnical and Tunnel Engineering, University of Innsbruck (geotechnik@uibk.ac.at) as Acrobat file (pdf) on demand

- Application of filters to remove spurious vectors, such as *SnR* (signal-to-noise ratio), the *global* and the *local filter*.
- Interrelation of the world coordinate system (WCS) with the images.
- Masking of images: To save computational time, regions of no interest can be masked out.
- Saving/loading of PIV calculations.
- Setting/saving/loading of calculation parameters (user-friendly environment).

Module II (Post-processor): For the graphical evaluation of the test results.

- Setting of the principal point.
- Loading and scanning of the specimen: definition of the region of interest[26] for the calculation of initial specimen diameter by comparing the gray scales line-by-line (minimization of the quadratic standard deviation, automatically done by the program).
- A graphical 3D representation of the calculated specimen deformations.
- Evaluation of triaxial test (stress-strain curve, volumetric strain) using PIV.

3.1.4.3   Restrictions of *SoilPIV*

Since the application of PIV to triaxial tests was used to prove the appropriateness of the shape assumption (Fig. 3.33 in section 3.2.1.3) for the evaluation of the triaxial tests (section 4.1), the use of local strain gauges (LVDTs), fixed diametrically on the specimen, was demanded. Therefore not the whole specimen was available for PIV calculations, because the LVDTs are visible on the acquired images (Fig. 3.11). This areas have to be avoided in the calculations. Only a sector[27] of the specimen was considered. The fixing of the LVDTs in the middle of the acquired image is also not preferable, because the cables would form an obstacle.

---

[26]The border of the triaxial specimen at the ends is not sharp when using lubricated ends, because it is not possible to suck the membrane that tight to the specimen without causing any initial consolidation. Therefore, masking out of top and bottom plate is necessary.

[27]This sector can be user-defined.

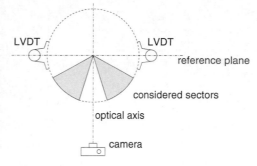

Fig. 3.11: LVDTs on the specimen visible on images for PIV calculations

Fig. 3.12: Top view of the discretization of the specimen for calculations with *SoilPIV*

Another problem arises from the (simple) mathematical models[28] 'real2virt' and 'virt2real' for calculating the specimen surface deformation, which has a singularity for points coinciding with the optical axis. Therefore, the sector in the middle of the above mentioned sector was avoided in further calculations, see Fig. 3.12. Due to the mathematical models to calculate 3D deformations from 2D images, small displacements close to the optical axis trigger large calculated radial deformations. The error-proneness is particularly high in this area.

3.1.4.4 Discussion of the results

Refraction parameters

The influence of the refraction indices on calculated distances of crosses on the reference plane, Fig. 3.8, in the water-filled triaxial cell was extensively investigated by Rinawi[18]. With the arbitrary variation of $n_p \pm 10\%$ the deviation compared to the value table 3.2 was ± 0.1 mm ($= 0.06\%$ of specimen diameter). The refraction index n_p of the used triaxial cell was not determined experimentally. Therefore, this numerical analysis was conducted, to assess the influence of a possibly bad chosen value. The influence of the air refraction parameter n_a due to variations in temperature ± 2 K, pressure ± 100 mbar, humidity $\pm 10\%$ or CO_2 ± 100 ppm is neglectible, since it influences n_a less than 10^{-4}. The refractive index of water changes with variations in pressure and temperature. Temperature changes $\pm 10°$C cause a change of n_w at the

---

[28]The simplifications are (i) optical axis and specimen center coincide and (ii) the radial deformation of the specimen is in every height circular, but its magnitude can be different in every considered vertical section.

third decimal and for pressures up to 3 MPa at the fourth decimal. But it turned out that even this small changes have a great influence on calculations and should, therefore, be considered in the mathematical models 'real2virt' and 'virt2real' (not done yet).

Parallax

Due to parallax the diameter of the specimen is slightly bigger, since the image is refered to a reference plane which lies behind the last visible point on the diameter, see Fig. 3.13.

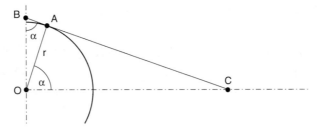

Fig. 3.13: Effect of parallax

The 'real' radius r of the specimen is:

$$r = c \cos \left(\tan^{-1} \frac{c}{y_B} \right) \quad , \tag{3.2}$$

being c the distance $\overline{OC}$ and y_B the distance $\overline{OB}$. The calculation of this 'real' radius is needed for the initial parameter set of PIV-calculations, which is evaluated from the undeformed specimen without water and triaxial cell. Therefore, refractions do not influence the calculations here.

Interrogation

MatPIV offersdifferent sizes of interrogation windows, which can be adopted to increase the density of the displacement field. Two methods are available to adopt the interrogation window size: (i) *singlepass* and (ii) *multipass*. The singlepass control is faster, but more susceptible to errors. It calculates the displacements from a single run. More reliable is the multipass routine, which halves the interrogation window sizes after each run and uses the last displacement field as prefered search direction for the new displacement calculation.

Due to the relatively simple mathematical models 'real2virt' and 'virt2real', the window overlap has to be set to 0.5 (50%). Window overlap is used to increase the number of displacement vectors. It is the distance of two adjacent interrogation windows. A value of 0.5 means that half of the first window is inside second window.

Accuracy of PIV

Errors can be quantified analysing noise-free simulated images by MatPIV, see e.g. DiFlorio *et al.*[29] With these images the exact pixel positions are known, errors solely arise from the processing by the program. With this knowledge, noise effects from the digital camera can be examined.

Okamoto *et al.*[30] distribute standard PIV images. With them the accuracy of the PIV analysis can be examined. They are available via the web-site `http://www.vsj.or.jp/piv`. These images contain 1-100,000 particles from tests with known velocities and can be user-defined downloaded there together with data files of particle informations. Fig. 3.14 contains two synthetic images. However, MatPIV is a program which is very widespread. For this stage of study it was assumed that sufficient accuracy is given by MatPIV – the core of *SoilPIV*.

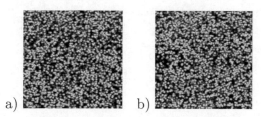

a)   b)

Fig. 3.14: Estimation of the accuracy with synthetic images; a) first, b) second image; downloaded from `http://www.vsj.or.jp/piv`

Submitted Data: Image Size 256×256 pixel, Actual Area 100×100 mm, Number of Particles 10000, Number of Images 2, Image Interval 100 msec, Laser light thickness 1 mm, Particle Size (Average) 12 pixel, Particle Size (Std. Deviation) 0 pixel, Particle Size (Minimum) 12 pixel

Camera alignment

A precise camera alignment is essential for reliable results using *SoilPIV*. The alignment has to be assured in three spatial directions. 'Perfect' alignment is achieved if the center of the camera lens is visible in a vertical aligned small mirror, fixed to the reference plane ("mirror trick").

---

[29]Di Florio, D., Di Felice, F., Romano, G.P.: Windowing, re-shaping and re-orientation interrogation windows in particle image velocimetry for the investigation of shear flows,*Measurement Science and Technology*, 7 (**13**) 2002, pp. 953-962

[30]Okamoto, K., Nishio, S., Saga, T., Kobayashi, T.: Standard images for particle image velocimetry, *Measurement Science and Technology*, 6 (**11**) 2000, pp. 685-691

If the camera is misaligned, all observed points refer to an unknown reference plane (Heikkilä and Silvén[31]), see Fig. 3.15. As can be seen from this figure, perspective projection is not shape preserving. Heikkilä and Silvén give equations for the correction of asymmetric projection. However, this correction was not done yet. The correction was replaced by the discretization (Fig. 3.12) and the assumptions taken for the routines 'virt2real' and 'real2virt', but is planned as matter of further research.

'Perfect' alignment was not achieved. The "mirror trick" was discovered after finishing the test series in Prague. All available test results contain errors due to misalignment, which manifest in different radial deformations calculated for the two considered sectors left and right of the optical axis (Fig. 3.16). This figure shows radial deformations resulting from conversion of displacement vectors onto the surface. The different calculated radial deformation developed although the velocity vectors on both sides of the optical axis possess the same absolute value. The influence of an artificial introduced horizontal eccentricity in the calculations was investigated in detail by Rinawi, showing that it was not possible to remove that error on the available tests. That fact was attributed to a more complex angular misalignment, not only horizontal. As can be seen from Fig. 3.16, the influence on the distortion of calculated radial deformations is more pronounced near the optical axis. Therefore, it was decided to cut out a central slice of 15° by default from the modelled specimen.

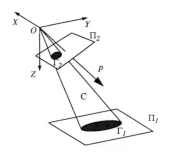
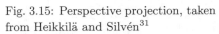
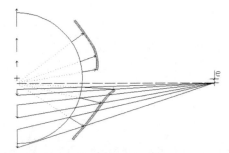

Fig. 3.15: Perspective projection, taken from Heikkilä and Silvén[31]

Fig. 3.16: Distortion of calculated radial deformation due to horizontal eccentricity e, taken from Rinawi

---

[31]Heikkilä, J., Silvén, O.: A four-step camera calibration procedure with implicit image correction, Proceedings of the IEEE Computer Society Conference on Computer Vision and Pattern Recognition (CVPR'97), San Juan, Puerto Rico, pp. 1106-1112

Comparison with local and global displacement measurements

A precise evaluation of triaxial tests using the PIV method in the here presented way was not possible due to (i) the relative simplicity of the mathematical models ('real2virt', 'virt2real') allowing the calculation of 3D deformations from 2D images and (ii) the poor alignment of the camera. Fig. 3.18 shows the stress-strain-curves, volumetric strains and deformed shape of a PIV-modelled triaxial test and the comparison with the results of this test, evaluated as described in section 3.2.1.3.

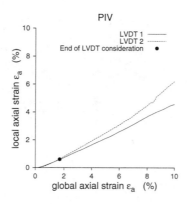

Fig. 3.17: Comparison of local (LVDT) and global axial strain measurements

(for further informations to this diagram see section 3.2.1.3)

The 'conventionally measured' stress-strain-curve Fig. 3.18 was calculated assuming a homogeneous deformation of the specimen until the bifurcation of the two local LVDT-measurements, see • in Fig. 3.17. After that point stress was calculated assuming barreling of the specimen. Further in Fig. 3.18 are plotted the minimum and maximum of 'PIV calculated' values, which correspond to the maximum and minimum cross section area of the specimen, evolved due to insufficient lubrication of the end platens. For this plotted test the specimen bulges relatively soon. Bulging can be seen in Fig. 3.17 from the bifurcation of the two local strain gauge measurements and from Fig. 3.18 in the difference in PIV-calculated stresses. The continuous line in Fig. 3.18 quite good converges to the minimum PIV-calculated values. Due to the aforementioned discrepancies of this first study of the PIV-method applied on triaxial tests[32], the reader is asked not to overrate the shown

---

[32]Discrepancies: uncertain estimation of initial specimen dimensions using PIV, uncertainty in estimation of refraction parameters, imprecision due to bad illumination, bad

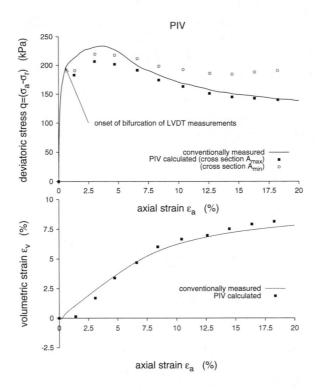

Fig. 3.18: Stress-strain-curve calculated from (i) area assumptions (section 3.2.1.3), (ii) PIV method

result. However, PIV was not available for all tests. But the appropriateness of the assumptions taken for the use of local and global measurements and specimen deformation (initially cylindrical, then barreling) has been shown.

Figures 3.19 and 3.20 show the PIV-calculated specimen during the triaxial test close to the peak and at the end of the test. Close to the peak no notably bulging is visible. The stress distribution over the height is plotted at the side. The color of the specimen does not have any importance.

alignement of the camera, and more

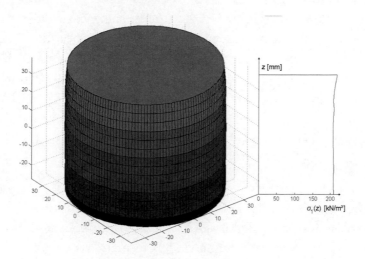

Fig. 3.19: Deformed (*SoilPIV*-modelled) specimen and stress distribution over height of the specimen near the peak of the stress-strain-curve

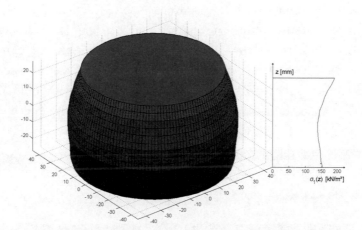

Fig. 3.20: Deformed (*SoilPIV*-modelled) specimen and stress distribution over height of the specimen at the end of the test

3.1.5 Specimen preparation

Laboratory tests with cohesionless sand have to be done with reconstituted specimens. Disturbing influences, such as inhomogeneities within the specimen — since they can trigger large errors in the measurements — have to be minimized. Uniform specimens (homogeneous density distribution within the specimens) are important for studies of soil mechanical behaviour, and one must be able to build replicable specimens for comparative reasons. The preparation of uniform specimens (minimization of inhomogeneities) is a quite difficult task. Several reseachers proposed methods to produce them, see later remarks. Inhomogeneities in density distribution can have many reasons, such as:

- segregation during specimen preparation,
- weaker zones, especially at the boundary of the specimen, when pouring the sand into a membrane lined split former,
- loosening due to leveling of the top surface.

The method to prepare the soil and soiltron specimens was spooning of five soil layers of known mass into a membrane lined split former with subsequent vibration of the mould under surcharge load on top of each layer. After leveling of the top surface, the top cap was placed onto the specimen and the membrane was sealed with O-rings. A small suction of 5 kPa was applied to the dry specimens to stiffen them and make them rigid enough to minimize disturbances due to the installation of the local gauges and the measurement of size. Specimens for tests with cell pressure lower than 1000 kPa were enclosed by one membrane, at higher pressures two membranes were required to prevent damage. Fig. 3.21 shows the preparation process for a test with sand.

Fig. 3.21: Specimen preparation steps

To avoid destruction of the membrane with sand particles, the surcharge had a slightly smaller diameter than the final specimen. However, this also may contribute to a smaller density at the specimen boundary.

3.1.5.1 Discussion of preparation methods for triaxial specimens

Pluviation

In recent literature pluviation is referred to as the most appropriate method to obtain uniform specimens, see e.g. Cresswell *et al.*[33] or Vaid *et al.*[34] This method is available for dry (air pluviation) and water saturated (water pluviation) specimens. With air pluviation, resulting densities can be controlled by varying the drop height and, hence, the impact velocity. Water pluviation can be used for the preparation of loose specimens.[35] There, the terminal velocity is reached already at a very small drop height, therefore it is not necessary to vary the drop height.

Water pluviation: A known weight of dry sand is boiled in a water filled flask and cooled down to room temperature. The the sand is deposited from the flask under water under gravitational influence into a de-aired water filled membrane lined split former. Densification can be achieved by vibrating under surcharge on the top cap[36] until a continuous controlled target height is reached.

Air pluviation: A known weight of dry sand is poored from a reservoir through a funnel and a series of diffuser meshes into a membrane lined split former. The drop height should be constant, therefore, the funnel is raised simultaneously with the deposit surface. Dry pluviated specimens require a long saturation time.

Within this study, dry specimens were investigated, therefore the following

[33] Cresswell, A., Barton, M.E., Brown, R.: Determining the maximum density of sands by pluviation, *Geotechnical Testing Journal*, 4 (**22**) 1999, pp. 324-328

[34] Vaid, Y.P., Negussey, D.: Preparation of reconstituted samples, in *Advanced Triaxial Testing of Soil and Rock*, ASTM STP 977, Donaghe, Chaney and Silver, Eds., ASTM Philadelphia 1988, pp. 405-417

[35] Denser saturated specimens can be achieved by water pluviation with subsequent vibration.

[36] after Vaid *et al.*[34]: leveling of the top surface after densification leads to a zone of smaller density on top, and should, therefore, be avoided

problems using air pluviation are only mentioned, for more information the reader is refered to Vaid et al.[34]:

- Well-graded soils segregate, because of different falling velocities.
- Higher pouring rates inhibit particles from acquiring a stable and dense configuration.
- Air may be entrapped in front of the sand rain. The achieved velocity, and, hence, density decreases.
- Frictional forces of the membrane lining in the split former inhibit denser packings.
- A precise control of drop height during pouring is necessary, because of the sensitivity of void ratios to falling velocity.

Moist tamping method

Sand of known weight and moisture content is placed in layers into a mould by spooning, and compacted. The water eases compaction. This method is often coupled with the so-called undercompaction method, where the layers get gradually denser with every layer added. The desired density is specified at the midheight of the specimen, the underlying layers are only compacted to a value by of, say, $-3\%\times$number of layers under the middle one. The overlying layers are compacted each, say, $+3\%$ of the previous layer density. This value is called the undercompaction ratio. The target density of each layer is controlled by dry weight and the layer size. Therefore, the height has to be controlled during the compaction process. A method to do so is fixing a stopper on the compaction rod, to inhibit further intrusion into the layer.

This method has the disadvantage, that the soil specimen is already some overconsolidated. X-ray analysis showed that a uniform distribution of density was not achieved with the moist tamping method, despite undercompaction (see Frost et al.[37] and Figures 1.27 and 1.28)

---

[37]Frost, J.D., Park, J.-Y.: A critical assessment of the moist tamping technique *Geotechnical Testing Journal*, 1 (**26**) 2003, pp. 57-70

3.1.5.2   Control of homogeneity

The density can be controlled with the following methods:

- Average density calculation from outer dimensions and weights of specimens.
- Grout in resin and cut in slices, analysed with image analysis.
- Analysis using X-rays, also with tomography.

Here, the first method was applied. In Fig. 3.22 is shown the average density for each of the five layers, calculated from the sum of layer weights and layer sizes. As can be seen from this figure, the void ratio distribution is not always homogeneous through the whole specimen. Different densities within the specimen enhance inhomogeneous deformations. In Fig. 3.22 are shown three different density distributions in the specimen:

a) A higher void ratio in lower layers.

b) A more or less homogeneous density distribution within the specimen

c) A higher void ratio in upper layers.

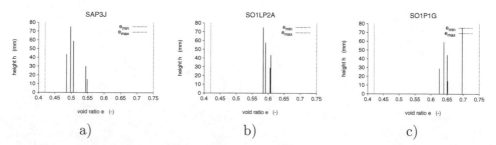

Fig. 3.22: Initial average void ratio of three specimens; a) higher void ratio in lower layers, b) quite homogeneous void ratio distribution within whole specimen and c) higher void ratio in upper layers

Although specimen preparation with spoon is not very reliable, it is the only applicable method for the present investigation, because soiltron consisting of sand and polystyrene or perlite is not possible to pluviate through air, due to different densities and subsequent falling velocities, and through water, due to floating of polystyrene and perlite.

For tests with saturated specimens to check the reliability of the volume change measurement using the U-tube and a GDS controller, the density was calculated from the dry weight of the specimen after the test and the initial size of the specimens.

3.1.6   Test types

To prove the assumption of the simulation of barotropy with pyknotropy, various triaxial tests were done with a quartz sand and sand mixed with two additives: polystyrene (Soiltron 1) and perlite (Soiltron 2). Sand and soiltron were investigated with drained triaxial tests with dry and saturated specimens. There were conducted tests with:

- isotropic consolidation: path OA
  These tests were done for the estimation of hypoplastic parameters h_s and n, see section 2.1.

- constant radial stress σ'_r: path OBC

- constant axial stress σ'_a: path OBE

- constant p': path OBF

- K_0-path: path OD

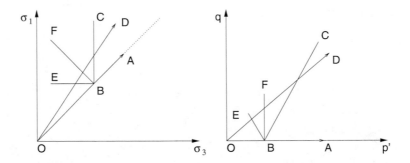

Fig. 3.23: Investigated stress paths

With the constant axial stress tests it was possible to test soiltron at very low pressures, the minimum cell pressure was 4 kPa. When working with this low pressures the control of cell pressure is very sensitive with GDS controllers, which have a resolution of ± 1 kPa, this would result in a large scatter in the obtained results. When starting from a higher cell pressure and keeping the subsequent initial axial stress during the test constant, the GDS controllers were more precisely to control.

3.2   Error control

Triaxial apparatuses are widely used to determine the stress-strain relations of soils in laboratory. Obtained results contain errors due to scatter and errors in measurements, uncertainties and variabilities of the soil properties, assumptions for the evaluation, etc. Typical sources of errors are (see i.e. Germaine and Ladd[38] or Baldi *et al.*[39]):

Systematic errors:		Random errors:

1. Sensors,	7. Frictional ends*,	1. Tilting/seating**,
2. Apparatus stiffness*,	8. Rate of loading*,	2. Saturation*,
3. Membrane stiffness,	9. Localization of deforma-	3. Temperature**,
4. Filter drain stiffness*,	tions,	4. Water leakage**,
5. Piston friction/uplift*,	10. Weight of top plate,	5. Gas leakage**.
6. Specimen geometry,	11. Compressibility of the load cell,	
	12. Membrane penetration.	

Items above marked with asterisk * have not been taken into account:

- Apparatus stiffness: The apparatus deforms under load. This leads to an error in measurement of axial deformation of the specimen when measuring only with external displacement gauges. Here, additional, internal LVDTs were used to measure the axial deformation of the specimens.

- Filter drain stiffness: Filter drains around the specimen were not used.

- Piston friction: The measurement of vertical load was measured inside the cell, therefore friction of the piston need not be considered.

- Uplift force must only be considered when using external load cells. It originates from the cell pressure, lifting the piston.

- Frictional ends stiffen the specimen, cause nonuniform deformation and trigger localization of deformation. Therefore, lubricated ends have been

---

[38] This section was inspired mainly by:
Germaine, J.T., Ladd, C.C.: Triaxial Testing of Saturated Cohesive Soils, in *Advanced Triaxial Testing of Soil and Rock*, ASTM STP 977, Donaghe, Chaney and Silver, Eds., ASTM Philadelphia 1988, pp. 421-459

[39] Baldi, G., Hight, D.W. and Thomas, G.E.: A Reevaluation of Conventional Triaxial Test Methods, in *Advanced Triaxial Testing of Soil and Rock*, ASTM STP 977, Donaghe, Chaney and Silver, Eds., ASTM Philadelphia 1988, pp. 219-263

chosen to improve the uniformity of deformation at all strain levels and to reduce the uncertainty in area correction.

— The rate of loading should be selected according to the time required for pore pressure dissipation throughout the specimen under drainage condition. Sands usually have a high permeability and thus the chosen loading velocity should not influence the pore pressure development. In the present case, the stress-strain behaviour of dry sand was investigated. The air permeability is much higher than the water permeability. All tests were carried out with a constant piston velocity.

— Saturation: The tests on sand and Soiltron were conducted with dry specimens.

Errors marked with double asterisk ** were avoided by proper execution of the tests (see section 3.2.2).

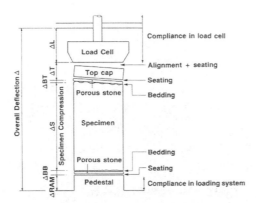

Fig. 3.24: Sources of errors in external axial deformation measurements, taken from Baldi *et al.*[39]

Fig. 3.25: Triaxial apparatus of the Soil Mechanics Laboratory at the Charles University in Prague

3.2.1 Systematic errors

3.2.1.1 Sensors

The sensors used in the tests were:
— Load cells: for the measurement of axial force,
— Linear variable differential transformer (LVDT): for the measurement of the axial displacement,

– GDS standard controllers: for the measurement and controlling of cell pressure and volume change.

The performance of a sensor is characterized by: stability (change in output at zero additional load or zero additional displacement over time), hysteresis and repeatability (variation of output for several loading cycles). All used sensors were carefully and repeatedly calibrated. Load cells were calibrated with weights (Fig. 3.26)and the LVDTs were calibrated using a micrometre (Fig. 3.4, p. 74).

Load cells

Data for the calibration were acquired for all load steps over a long period of time to cover the whole bandwidth of measured data and to check the stability, as can be seen exemplary in Fig. 3.27. The maximum scatter was about $\pm$ 17 units, with a range of 65536 units (about $\pm$ 0.03%).

Hysteresis was analysed with a loading and unloading cycle. No mentionable differences of the measured values was detected. The calibration data satisfy linear regression with respect to the applied load.

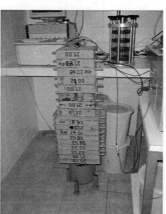

Fig. 3.26: Calibration of the load cell

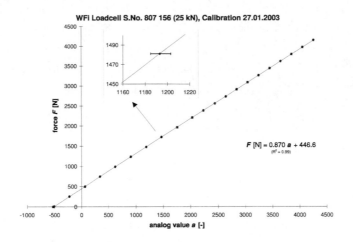

Fig. 3.27: Result of the calibration of the load cell

LVDTs

As with load cells, hysteresis, stability and repeatability were checked. A typical result of calibration is shown in Fig. 3.28. The maximum deviation during one calibration cycle (=stability) was ± 30 units (± 0.05%, comparable to specification of manufacturer). Again, no mentionable difference was measureable during unloading.

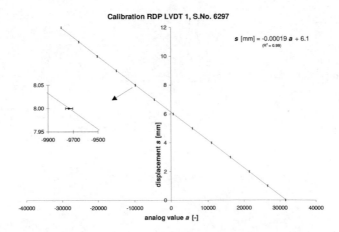

Fig. 3.28: Result of the LVDT calibration

3.2.1.2 Membrane stiffness

The rubber membrane restrains specimen deformation. The corrections are based on mebrane theory and assume elastic material, continuous support without buckling and uniform cylinder deformation (for more information the reader is referred to Germaine and Ladd.[40]) Corrections are expressed in terms of an additional axial force and radial stress, respectively. The following approximations were suggested by BERRE (cited in Germaine and Ladd[40]):

$$\Delta F_a = -\pi D_{im} t E \left(\varepsilon_a + \frac{2}{3} \varepsilon_v \right) \tag{3.3}$$

$$\Delta \sigma_r = 2tE \left(\frac{D_i - D_{im}}{D_i D_{im}} \right) - \frac{4tE}{3D_i} \varepsilon_v \tag{3.4}$$

with:

t initial membrane thickness

E YOUNG's modulus of rubber (typical values
1100 kN/m^2 (Molenkamp/Luger[41]), 1300 kN/m^2 (Baldi/Nova[42]),
1195 kN/m^2 (Ali *et al.*[43]), 1400 kN/m^2 (Germaine/Ladd[40])

D_i initial specimen diameter

D_{im} initial unstressed membrane diameter

ε_a axial strain

ε_v volumetric strain of the specimen

In Fig. 3.29 one can see that the consideration of membrane stiffness leads to a small reduction in axial stress. Calculations were done with D_{im}=73 mm, t=0.3 mm and E=1400 kN/m^2, which led to a maximum deviation of the deviatoric stress of $\approx$0.3% (Fig. 3.30, right). The correction depends in the case of test SAP2F (Fig. 3.29) mainly on axial strain, the volumetric strain is relatively small (Fig. 3.30, left). The correction increases more or less linear.

---

[40] Germaine, J.T., Ladd, C.C.: Triaxial Testing of Saturated Cohesive Soils, in *Advanced Triaxial Testing of Soil and Rock*, ASTM STP 977, Donaghe, Chaney and Silver, Eds., ASTM Philadelphia 1988, pp. 421-459

[41] Molenkamp, F., Luger, H.J.: Modelling and minimization of membrane penetration effects in tests on granular soils, *Géotechnique*, 4 (**31**) 1981, pp. 471-486

[42] Baldi, G., Nova, R.: Membrane penetration effects in triaxial testing, *Journal of Geotechnical Engineering*, 3 (**110**) 1984, pp. 403-420

[43] Ali, S.R., Pyrah, I.C., Anderson, W.F.: A novel technique for the evaluation of membrane penetration, *Géotechnique* 3 (**45**) 1995, pp. 545-548

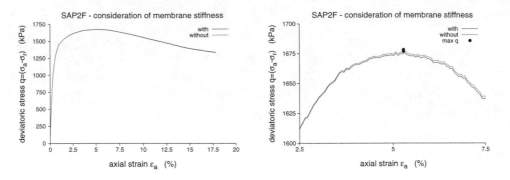

Fig. 3.29: Results of a typical test, taken membrane correction into account; right: zoomed

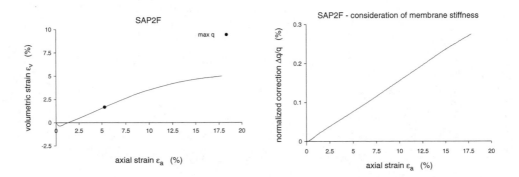

Fig. 3.30: Volumetric strain (test SAP2F); Normalized membrane correction

3.2.1.3   Area corrections

In triaxial tests, the specimen cross section area — and therefore the vertical stress — are usually computed with the initial cross section and the axial and volumetric strains, assuming a homogeneous cylinder deformation of the specimen:

$$A_c = A_0 \left(\frac{1 - \varepsilon_v}{1 - \varepsilon_a} \right) \tag{3.5}$$

where A_0 denotes the cross section area corresponding to zero strain.

For lubricated end platens the assumption of uniform deformation is acceptable for consolidation and the initial stage of shearing. Bulging and barreling of the specimen induce inhomogeneous deformations with variable specimen diameter and thus equ. (3.5) leads to an overprediction of axial stress (see also Fig. 3.34).

In case of a nonuniform deformation the corrected cross section area at the

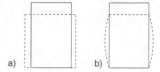

Fig. 3.31: Shapes of specimens after Germaine/Ladd[40]; a) Cylinder, b) Parabolic (barreling or bulging)

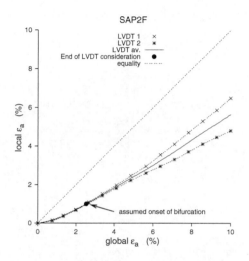

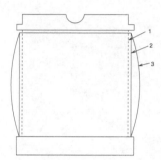

Fig. 3.32: Consideration of local axial deformation measurements (test SAP2F). The bifurcation indicates the onset of inhomogeneous deformation.

Fig. 3.33: Assumption of specimen deformation: 1 homogeneous cylinder deformation, 2 specimen at the end of the homogeneous deformation, 3 bulging

largest midplane section can be computed, assuming a constant diameter of the specimen at the specimen ends and a parabolic shape of the meridian of the specimen, after Germaine/Ladd[40]:

$$A_c = A_0 \left(-\frac{1}{4} + \frac{\sqrt{25 - 20\varepsilon_a - 5\varepsilon_a^2}}{4\left(1 - \varepsilon_a\right)} \right)^2 \tag{3.6}$$

The specimen deformations in the conducted tests were almost always a mixture of the aforementioned deformation modes (see Fig. 3.31). Thus area corrections are needed. Note, that all applied area corrections are only approximative due to uncertainty in the measurement of the specimen shape, which could be checked by measurement of the specimen's dimensions after removal of the cell pressure and dismounting the cell.

Results were evaluated on the basis of LVDTs. Measurement up to the onset of barreling, see e.g. point •, Fig. 3.32, and subsequently from the global deformation measurement, assuming barreling of the specimens, Fig. 3.33 and using equ. (3.6).

Fig. 3.32 shows raw data of the global and local deformation measurements. As can be seen, the global strain is higher in the beginning than the local one, due to squeeze out of the lubrication grease and smoothing of the specimen surfaces adjacent to the end platens. It was assumed that this bedding error gradually disappeared. Subsequent strain and axial stresses are evaluated from the global deformation gauge, taking into account barreling of the specimens.

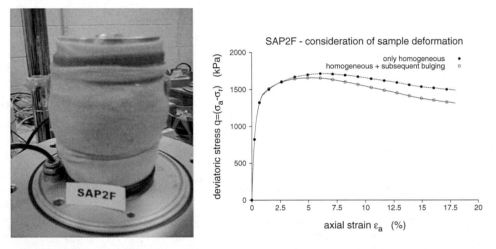

Fig. 3.34: Consideration of the deformation shape (test SAP2F); Application of area correction

3.2.1.4   Shear localization

Localization of deformations appears in compressed triaxial specimens with frictional ends. It occurs close to the peak (e.g. Tatsuoka et $al.$[44]), therefore tests with a drastic decrease of strength were considered only up to the peak or a drastic drop of axial stress, or they were discarded (see Fig. 3.35).

3.2.1.5   Weight of top plate

The loading of the specimen due to the top plate can be critical for tests with low cell pressures. For our tests, a polished top plate (Fig. 3.36) and

---

[44]Tatsuoka, F., Nakamura, S., Huang, C., Tani, K.: Strength Anisotropy and Shear Band Direction in Plane Strain Tests of Sand, $Soils$ and $Foundations$, 1 (**30**) 1990, pp. 35-54

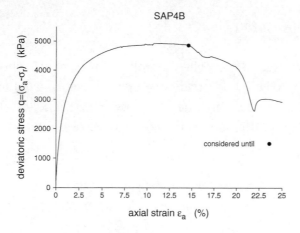

Fig. 3.35: Localisation in triaxial test (SAP4B)

lubrication was used. The minimal initial cell pressure due to the cell liquid was approximately 2 kPa. Therefore, the weight of the top plate, see table 3.3, should not influence the specimen.

plate m_{plate} (g)	plate and lubrication layers $m_{plate+lubrication}$ (g)	area pressure p (kPa)
231.32	234.64	0.521

Tab. 3.3: Top plates for the triaxial tests in Prague

Fig. 3.36: End plates, used for the triaxial tests

3.2.1.6   System rigidity

The rigidity was tested in a so-called compliance test. A stiff specimen-dummy is placed into the triaxial cell and is compressed as a usual specimen. The measured deformation contains the compression of the load cell, the bedding error and the deformation due to the compliance of the loading system.

Compressibility of the load cell

The load cell is deformable. It converts a load acting on it into electrical signals, because electrical resistance changes when the gauges deforms.

Bedding errors

The application of lubricated end platens cause bedding errors (Fig. 3.24 and 3.37) mainly due to the compression of the membrane sheets and silicone grease (Fig. 3.2) during loading. Additionaly, a contact of a granulate and platens results in a zone of higher porosity and thus higher compressibility (Fig. 3.37). Inclined specimen ends trigger excentric loading, another source of bedding error.

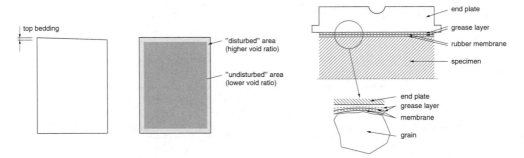

Fig. 3.37: Sources of the bedding error; cross section of a specimen: top bedding (inclined specimen surface); different void ratio distribution within specimen; indentation of grains into the lubrication layers

To evaluate the influence of lubrication, tests with and without application of lubricated ends were conducted (Fig. 3.38). The difference of the displacements at the same axial force can be expected to show approximately the magnitude of bedding error. Its amount is ca 1.2 mm, which roughly corresponds with the thickness of the lubricating layers (4×membranes of 0.3 mm + grease, thickness not measureable with the available means). Note, however, that the difference in these two tests do not show the *'true'* bedding error of sand or Soiltron specimens. When working with sand or Soiltron, the specimen surfaces adjacent to the lubricated ends are not as smooth as the dummy surface due to the grain edges and corners penetrating more easily into the lubricating layer (fig 3.37). Therefore the magnitude of bedding error in the tests with granular materials should be slightly larger than with dummy specimens.

Compliance of the loading system

As shown in Fig. 3.38, the stress-strain curve of the steel dummy exhibits initially a low stiffness, even without lubrication. This can only be attributed to the seating error. Under this assumption the obtained curves are rectified, i.e. they are shifted towards the origin by $\Delta u = 0.4 \ mm$. This part is as-

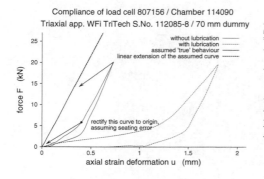 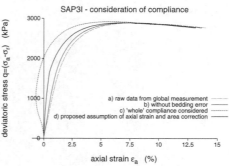

Fig. 3.38: Compliance of the loading system   Fig. 3.39: Effect of the consideration of compliance, bedding errors and compressibility of the load cell on the results of the triaxial tests

sumed as the compressibility of only the load cell. Of course, this is only a simplification, but this decision must be implemented, to make sure that the results remain feasible (the axial strain e.g. would be negative, when applying the 'whole' measured compliance, see Fig. 3.39, where different possible considerations of compliance, compressibility of the load cell and bedding errors are confronted).

The corrected axial strain with consideration of the 'whole' compliance of the system is initially negative, see Fig. 3.39 curve c). This is not plausible. Therefore, this approach seems not reliable and was not taken into account. Curve b) in this figure shows the results from the global axial deformation measurement of the test when lubrication and the initial not-assessable bedding error are not considered. Here also a reduction in the calculated axial

strain is visible. But still this cannot be the 'true' deforation of the specimens due to bedding errors, compliance and compressibility in the load cell.
Therefore, the assumed result of this test can be seen in curve d) of this figure,
considering

- initially homogeneous deformation, measured locally with the LVDTs
  and
- secondary bulging, deformation measured globally, assuming the decay
  of bedding errors (see shifted curve in Fig. 3.38).

3.2.1.7   Membrane penetration

Fig. 3.40: Membrane penetration in a triaxial test. Due to the cell pressure the membrane
is pressed into the interstices of the soil skeleton.

Due to the confining pressure the rubber membrane, as can be seen in Fig. 3.40,
is pressed into the interstices of the soil skeleton. Conventional drained
isotropic consolidated triaxial tests (CID) are evaluated starting from the
end of the isotropic consolidation. To calculate the volume of the specimen
after consolidation V_c, it is usually assumed that:

$$V_c = V_0 - \Delta V \quad ,$$

where V_0 is the initial volume, calculated from the dimensions of the specimen and ΔV is the volume change due to consolidation. Assuming isotropic

behaviour, one can calculate height h_c and diameter d_c after consolidation from:

$$h_c = h_0 \cdot \sqrt[3]{V_c/V_0}$$
$$d_c = d_0 \cdot \sqrt[3]{V_c/V_0} \;,$$

with h_0, d_0, V_0 being the initial dimensions of the specimen before consolidation and V_c after consolidation. These calculations are inaccurate since the measured volume change is strongly influenced from the membrane penetrating into the specimen during consolidation. The measured ΔV during consolidation is too high. Molenkamp/Luger[45] wrote:

> ...in drained isotropic loading and unloading the error in volume change can be of the order of 20-40%...

Kiekbusch/Schuppener[46] defined the unit membrane penetration ν_{mp} as:

$$\nu_{mp} = \frac{\varepsilon_{v,mp}}{A_m} \;, \tag{3.7}$$

with $\varepsilon_{v,mp}$ being the part of volumetric strain due to membrane penetration and A_m the surface area of the confining membrane.

Factors affecting membrane penetration

The membrane penetration is mainly affected by grain size, confining pressure and stiffness of the membrane. As representative size of the grains, the grain size d_{50} can be used. The particle shape and specimen density seem to be of minor influence (Baldi/Nova[47]).

To relate confining pressure to membrane penetration, often diagrams are plotted: a) $\varepsilon_{v,mp}$ vs. $\log \sigma_3'$ or b) $\varepsilon_{v,mp}$ vs. A_m (area of contact between membrane and specimen, to show also the effect of specimen geometry). From that, several empirical relations containing factors for grain size d_{50}, confining pressure σ_3', stiffness and thickness of the membrane E, t and dimensions of specimen d are being developed.

$$\varepsilon_{v,mp} = f(d_{50}, \sigma_3', d, E, t,)$$

[45]Molenkamp, F., Luger, H.J.: Modelling and minimization of membrane penetration effects in tests on granular soils, *Géotechnique*, 4 (**31**) 1981, pp. 471-486

[46]Kiekbusch, M., Schuppener, B.: Membrane penetration and its effect on pore pressures, *Journal of the Geotechnical Engineering Division*, 11 (**103**) 1977, pp. 1267-1279

[47]Baldi, G., Nova, R.: Membrane penetration effects in triaxial testing, *Journal of Geotechnical Engineering*, 3 (**110**) 1984, pp. 403-420

With increasing diameter of the specimen the susceptibility of membrane penetration decreases.

Newland/Allely[48] were the first to investigate the membrane penetration in triaxial tests. Since they assumed isotropic deformation ($\varepsilon_r = \varepsilon_a$, $\varepsilon_v = 2\varepsilon_r + \varepsilon_a$) of a sand specimen under hydrostatic loading, the membrane penetration can be calculated from the measured volumetric change and the axial deformations as

$$\varepsilon_{v,mp} = \varepsilon_v - 3\varepsilon_a \ \ .$$

Due to anisotropy the accuracy of this approach is questionable.

Laboratory methods

Raju et al.[49] improved the proposed method by Roscoe et al.[50] In both methods the interior of the soil specimen is partly replaced with brass dummy rods of various diameters (Fig. 3.41a). The membrane penetration is calculated assuming that the soil volume decreases, while the surface of the specimen exposed to membrane penetration remains the same. Roscoe et al. used a stiff top platen for their tests, whereas Raju et al. used a flexible top platen with a hole of the same diameter as the brass dummy and free to move on the dummy, to fulfil the requirement of isotropic loading of the soil specimen. The plot of volume change due to isotropic loading over the dummy diameter yields the volume change due to membrane penetration.

Kiekbusch et al.[51] equipped a triaxial cell with a circular specimen inside a base plate with a recess and covered it with a membrane (Fig. 3.41b). The specimen was fully saturated and pressurized. The vertical deformation of the specimen and the expelled water were measured, the difference gave the membrane penetration.

Lade et al.[52] investigated the membrane penetration using small brass plates between soil and membrane and comparing the volumetric behaviour of tests without using these plates.

---

[48]Newland ,P.L., Allely, B.H.: Volume changes in drained triaxial tests on granular materials, *Géotechnique*, 1 (**7**) 1957, pp. 17-34

[49]Raju, V.S., Sadasivan, S.K.: Membrane penetration in triaxial tests on sands, *Journal of the Geotechnical Engineering Division*, 4 (**100**) 1974, pp. 482-489

[50]Roscoe,K.H., Schofield, A.N., Thurairajah, A.: An evaluation of test data for selecting a yield criterion for soils, *ASTM Special Technical Publication No. 361*, 1963, pp. 111-128

[51]Kiekbusch, M., Schuppener, B.: Membrane penetration and its effect on pore pressures, *Journal of the Geotechnical Engineering Division*, 11 (**103**) 1977, pp. 1267-1279

[52]Lade, P.V., Hernandez, S.B.: Membrane penetration effects in undrained tests, *Journal of the Geotechnical Engineering Division*, 2 (**103**) 1977, pp. 109-125

Baldi *et al.*[47] measured the membrane penetration in drained triaxial tests as the difference between total volume change (measured with a burette), the axial deformation (measured with a LVDT between top cap of the specimen and cell) and the lateral deformation (measured with a LVDT).

Ali *et al.*[53] proposed to cement the soil specimen to make it sufficiently rigid to be assumed to have zero volume change under loading. Then the measured volume change must be only due to membrane penetration.

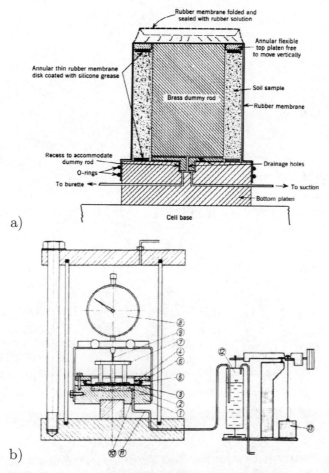

Fig. 3.41: Experimental setups from literature to assess membrane penetration, a) Raju *et al.*[49], b) Kiekbusch *et al.*[51]

[53] Ali, S.R., Pyrah, I.C., Anderson, W.F.: A novel technique for the evaluation of membrane penetration, *Géotechnique* 3 (**45**) 1995, pp. 545-548

Theoretical approaches to relate membrane penetration to cell pressure are, e.g.[54]:

$$\varepsilon_{v,mp} = n \left(\frac{d_{50}}{d} \right) \left(\frac{d_{50}}{Et} \right)^{1/3} \sigma_3'^{\,1/3} \ , \tag{3.8}$$

resp. Ramana & Raju[55] for d_{50}=0.08-2.00 mm:

$$\varepsilon_{v,mp} = 0.05 \frac{d_{50}}{d} \log \left(\frac{\sigma_3'}{\sigma_{3i}'} \right) \ , \tag{3.9}$$

with:

$\varepsilon_{v,mp}$ volumetric strain due to membrane penetration

d_{50} mean grain size

d specimen diameter

E Young's modulus of the membrane
(typical values for latex rubber[56,57,58]: range from 1,100-1,300 kN/m^2)

t thickness of the membrane

σ_3' effective lateral stress

σ_{3i}' initial σ_3', for membrane penetration consideration after consolidation

n Baldi & Nova[57]:   0.50

       Molenkamp & Luger[58]:   0.64

       Kramer *et al.*[59] *simplified*:   0.924

$$improved: \quad 1.58 \left(\frac{1-\alpha}{5 + 64\alpha^2 + 80\alpha^4} \right)^{1/3} \ with$$

$$\alpha = 0.15 \left(\frac{d_{50}}{Et} \right)^{0.34} p^{0.34} \ .$$

---

[54]used variables adjusted to unify the declarations

[55]Ramana, K.V., Raju, V.S.: Membrane penetration in triaxial tests, *Journal of the Geotechnical Engineering Division*, 2 (**108**) 1982, pp. 305-310

[56]Ali, S.R., Pyrah, I.C., Anderson, W.F.: A novel technique for the evaluation of membrane penetration, *Géotechnique* 3 (**45**) 1995, pp. 545-548

[57]Baldi, G., Nova, R.: Membrane penetration effects in triaxial testing, *Journal of Geotechnical Engineering*, 3 (**110**) 1984, pp. 403-420

[58]Molenkamp, F., Luger, H.J.: Modelling and minimization of membrane penetration effects in tests on granular soils, *Géotechnique*, 4 (**31**) 1981, pp. 471-486

[59]Kramer, S.L., Sivaneswaran, N.: Stress-path dependent correction for membrane penetration, *Journal of Geotechnical Engineering*, 12 (**115**) 1989, pp. 1787-1804

Laboratory tests to assess membrane penetration for Ottendorf-Okrilla sand

In this study a method based on Boháč/Feda[60] was used. Oedometric compression of a soil specimen in the triaxial cell was achieved through adjustment of cell pressure. Axial (from global deformation measurement) and volumetric strain (from the expelled drainage water) were measured independently. The radial stress was increased whenever a change in diameter of max. 0.02 mm occured, to keep $\varepsilon_r \approx 0$ (Fig. 3.42), which is boundary condition for oedometric tests. Therefore, membrane penetration can be calculated from

$$\varepsilon_{v,mp} = \varepsilon_v - \varepsilon_a \qquad (3.10)$$

The advantage of this method is, that one could use the same size of specimen and membranes as for shear strength test for the soil (but was not realized in this study, because a 75 mm strain belt was not available).

A saturated sand specimen was built up on the pedestal of a triaxial cell within a formwork and a radial strain belt with a LVDT, which was placed on it in midheight. The back and cell pressure were simultaneously increased to 10 kPa, to stabilize the specimen and the pressure/volume change measurement system. Then the specimens were consolidated with differential pressure 10 kPa.

Test No.	e_{0i}	No. of membranes	initial area of membrane A_{mi} [mm^2]
1.2	0.463	1	9659.6
1.5	0.426	1	9988.7
1.6	0.403	1	9916.8
1.7	0.430	1	10005.1

Tab. 3.4: Tests to assess magnitude of membrane penetration to Ottendorf-Okrilla sand

---

[60]Boháč, J., Feda, J.: Membrane penetration in triaxial tests, *Geotechnical Testing Journal*, 3 (**15**) 1992, pp. 288-294

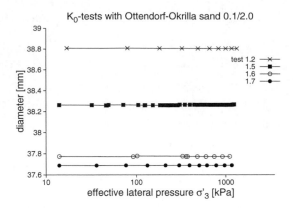

Fig. 3.42: Change of diameter during the tests

The parasitic volumetric strain due to membrane penetration was measured from:

- cell pressure
- drainage pressure and drainage volume change
- axial deformation of the specimens from an external dial gauge
- radial deformation of the specimens with local radial mounted LVDT

The external measurement of ε_a contains also the compliance of the load system and bedding errors (see also section 3.2.1.6). Therefore the presented results are only approximative. These tests were done to check the validity of the proposed equations (3.8) and (3.9) from the literature for high stresses, because the aforementioned authors did their tests only up to approximately 600 kPa. Within this study, the mechanical properties of sand were investigated in triaxial tests up to 2000 kPa.

In these tests the specimens were surrounded by one membrane. Fig. 3.43 shows the measured values of volumetric and axial strains. The difference of these two values (region above the dotted line) is the parasitic (spurious) volumetric strain due to membrane penetration, also shown in Fig. 3.44, where in addition the approximations equ. (3.8) are diagrammed.

Conclusion

The test results and equations (3.8) and (3.9) match quite good. Therefore it was decided to use equ. (3.8) with the $n = 0.64$ (as proposed by Molenkamp/Luger[58], which is approximately the average of the membrane

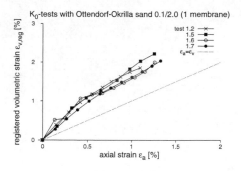

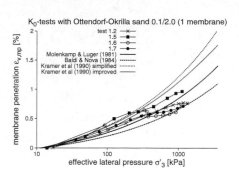

Fig. 3.43: Registered volumetric strain $\varepsilon_{v,reg}$ vs. measured axial strain ε_a

Fig. 3.44: Volumetric strain due to membrane penetration $\varepsilon_{v,mp}$ vs. effective lateral stress σ'_3, comparison with empirical solutions

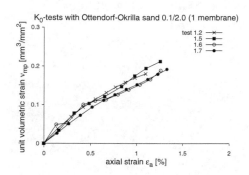

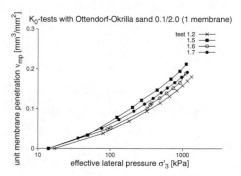

Fig. 3.45: Unit volumetric strain, equ. (3.7), due to membrane penetration $\nu_{v,mp}$ vs. axial strain ε_a or efective lateral stress σ'_3, respectively

penetration) for the evaluation of the specimen volume after consolidation.

$$\varepsilon_{v,mp} = 0.64 \left(\frac{d_{50}}{d} \right) \left(\frac{d_{50}}{Et} \right)^{1/3} \sigma'^{1/3}_3 \qquad (3.11)$$

During the test the membrane penetration was assumed as constant if the cell pressure was kept constant. Fig. 3.45 shows the unit volumetric strain, see eqn. (3.7), due to membrane penetration.

3.2.2 Random errors

3.2.2.1 Seating and tilting

The presented tests were done with reconstituted specimens. The specimens were built up on the pedestal of a triaxial cell in a mould. During preparation of the specimens a centered mould was positioned on the base platen in order to minimize tilting of the specimen. The specimen's upper surface was carefully smoothed to minimize the bedding error.

The error in tilting can be estimated by measuring the specimen height on several points. The maximum difference was 1.5 mm (Test SAP2G) $\rightarrow \Delta h/\varnothing = 2\%$. However, after lowering the piston, an uneven contact between the piston and the upper plate can produce a slight inhomogeneous deformation. This movement and the corresponding tilt cannot be measured. Tilting of a specimen can occur also during a test (see Fig. 3.46) due to inhomogeneous specimen deformation and is not taken into account.

Fig. 3.46: Tilting of a specimen; during and after the test (SAP2b)

Seating errors can occur due to nonparallel surfaces between the internal load cell and the drainage plates (see also Fig. 3.24). In the presented case, a thin porous stone was inserted into a recess of the bottom plate (see Fig. 3.36) to avoid seating error.

3.2.2.2 Temperature

Temperature changes were inevitable, but were minimized, because the laboratory is situated in the basement and the duration of the tests with sand and

Soiltron was quite short ($\approx$ 2 h). The maximum absolute temperature change during a triaxial test was observed as 1.2 K (test SO1P3C). The influence on the test results can be assumed as neglible

3.2.2.3   Water leakage

Water leakage can occur at the sealing of the cell and valves (external leakage) or into the specimen (internal leakage). Leakage at the sealings of the cell was not a problem as long as the cell pressure was kept constant. Cell liquid volume changes were not considered in any calculations. The volume changes of the specimens during tests were calculated from the air expelled from or drawn-in the specimen. Leakage into the specimen was observed in a few tests by colouring the specimen and run out of water through the air drainage. These tests were interrupted.

3.2.2.4   Gas leakage

Gas can in waterdrained triaxial tests diffuse from the cell water through the membrane into the specimen's drainage water and, therefore, falsify the measured amount of expelled or drawn-in drainage liquid. This can be avoided by backpressurizing the specimen. However, gas leakage becomes serious in long-term experiments only with saturated specimens. Here, dry specimens were investigated in short term tests, hence this error did not occur.

3.3   Reliability of test results

Soil properties are often afflicted with uncertainties, caused by natural variability (inherent physical randomness) scatter in acquired data due to soil variability and errors in the measurements, and lack of knowledge of the mechanical behaviour. Due to the variations in mineral composition and stress history in the soil mass, it is difficult to assign a single soil parameter to a soil mass from spot checks. Soil sampling is often afflicted with errors, caused by specimen disturbances, test imperfections and human factors. Reliable results can be attained by taking more specimens and conduct more tests. However, for scientific investigations the accuracy must be higher than for general engineering purposes.

Observations (results of measurements: instrument readings, conversions to engineering parameters, evaluations, etc.) exhibit always scatter, which tends to exhibit regularity if the number of observations is large. One possibility to reduce the variations is the repitition of tests. But is after Baecher and Christian[61] unnecessary, since normal sampling variation already leads to high differences in the results.

In the following, the scatter caused by the measurement devices and the consequences on the results of the triaxial tests with Ottendorf-Okrilla sand and Soiltron are analysed.

3.3.1   Soil behaviour obtained from laboratory tests

How can we judge the reliability of results? In the here presented triaxial tests various measurement techniques were used. Let's have a closer look for example at the deformation measurement.

Following strain gauges were used[62]:

- Externally mounted: Used for the *'global'* measurement of axial specimen deformation, which is equal to the displacement of the triaxial cell, if bedding errors and other parasitic axial displacement measurements are neglected.

---

[61] Baecher, G.B., Christian, J.T.: Reliability and Statistics in Geotechnical Engineering, John Wiley & Sons Ltd., Chichester, 2003

[62] Gauges and errors described in detail in sectionss 3.1 and 3.2.

– Mounted on the specimen: Used for the *'local'* measurement of axial specimen deformation, which is assumed as equal to the displacements of the fixings, without any parasitic deformations. Usually, two diametrical opposite displacement transducers are used.

With these three different strain gauges we get three different results, see e.g. Fig. 3.47, where the deviatoric stress and volumetric strain over the axial strain of a triaxial test on sand are plotted. The results of peak friction angle and peak axial strain are summarised in tab. 3.5.

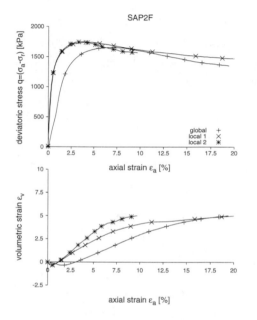

Fig. 3.48: Specimen SAP2F after the test. The deformation pattern is like a barrel. Note that the local strain gauges were already dismantled before taking this photo.

Fig. 3.47: One test – three results

	global	local 1	local 2
$\varepsilon_{a,Peak}$	7.1%	4.4%	3.7%
φ_{Peak}	38.6°	39.4°	39.5°

Tab. 3.5: Friction angles and peak axial strains of test SAP2F obtained with various methods

The global measurement is the most unreliable one due to bedding error from the lubrication of the specimen ends. The local measurements are not reliable in a later stage of the test due to barreling of the specimen, see Fig. 3.48.

To obtain e.g. the peak friction angle, we have to take assumptions and simplifications for the evaluations. One assumption is the *uniform stress distribution* within the specimen which can at best be achieved when an uniform deformation field through the specimen (homogeneous axial strains) is fulfilled. It is known that a combination of rubber sheeting and silicone grease reduce the friction on the end platens and avoid the formation of quasi-rigid zones at the specimen ends. So-called bedding errors arise due to squeezing out of lubrication grease and the compressibility of the rubber membranes. Therefore, deformation has to be monitored locally with linear variable differential transformers (LVDTs) on the specimen or with remote devices such as optical or x-ray techniques.

Local strain gauges have to be mounted to the specimen, but care must be taken not to influence the specimen response through stiffening the specimen with the mounted equipment. However, LVDTs have many advantages[63]:

- – nearly frictionless measurement,
- – long mechanical lifetime,
- – extremely high resolution, and
- – core and coil separation (eases mounting).

and were therefore chosen for the triaxial testing program. The appropriateness of the LVDT-measurements, assumptions of specimen deformation and utilisation of global deformation measurements was checked with remote sensing using digital imaging and evaluation with Particle Image Velocimetry (PIV) as described in section 3.1.

3.3.2   Post-peak behaviour

Can we trust results beyond the peak? In standard triaxial tests deformation localises into shear bands. Fig. 3.49 shows a computer tomography image of a triaxial test specimen. The localisation was e.g. investigated by Alshibli *et al.*[64] aboard the Space Shuttle in zero gravity. It appears that the

[63]Linton, P.F., McVay, M.C., Bloomquist, D.: Measurements of Deformations in the Standard Triaxial Environment with a Comparison of Local versus Global Measurements on a Fine, Fully Drained Sand, in *Advanced Triaxial Testing of Soil and Rock*, ASTM STP 977, Donaghe, Chaney and Silver, Eds., ASTM Philadelphia 1988, pp. 202-215

[64]Alshibli, K.A., Sture, S., Costes, N.C., Frank, M., Lankton, M., Batiste, S., Swanson, R.: Assessment of Localized Deformations in Sand Using x-ray Computed Tomography, *Geotechnical Testing Journal*, ASTM, 3 (**23**) 2000, pp. 274-299

deformation patterns and localisations are dependent on gravity and confining pressure. In the figures shown here, one can see radial shear bands concentrated in the mid-height of the specimen. Under earth's gravity this symmetrical shear banding does usually not occur. E.g. Desrues and Viggiani[65] report on biaxial tests on Hostun sand which were observed with stereo-photogrammetry. The tests showed diverse localisation patterns although much effort was put on the preparation of the specimens.

Fig. 3.49: Localizations in triaxial specimen made visible using Using CT 3D Rendering

(taken from http://www.cee.lsu.edu/~ceegeotec/archives_micro.html)

If shear banding occurs, the axial deformation and volume change measurements refer to rigid body movement(s). In the shear zone the material becomes softer. Shear banding can be observed in triaxial tests when the measured volume change is continuously rising. The specific volume of the soil specimen in laboratory tests is usually averaged over the whole specimen size. Desrues *et al.*[66] observed triaxial sand specimens in triaxial compression with computed tomography. They measured within the shear bands a limit void ratio. This void ratio is higher than the global measured void ratio, which is an average value for the whole specimen.

For the triaxial tests of section 4.1 lubricated end platens were used to minimize the restraints of the specimens at the ends, therefore no well-defined shear band were observed through the membrane.

---

[65]Desrues, J., Viggiani, G.: Strain localization in sand: an overview of the experimental results obtained in Grenoble using stereophotogrammetry, *International Journal for Numerical and Analytical Methods in Geomechanics*, 4 (**28**) 2004, pp. 279-321

[66]Desrues, J., Chambon, R., Mokni, M., Mazerolle, F.: Void ratio evolution inside shear bands in triaxial sand specimens studied by computed tomography, *Géotechnique*, 3 (**46**) 1996, pp. 529-546

3.3.3   Data scatter due to measurement devices

The measurements were acquired by a computer in short lags. Therefore, a huge amount of data was collected, so that scatter in measurement device almost vanish, see fig. 3.50.

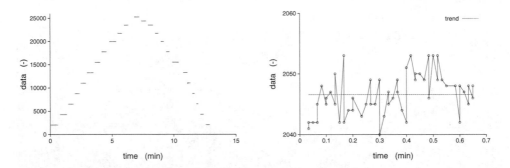

Fig. 3.50: Scatter of the load cell signal during the calibration process (done on load cell: s.nr. 22 843 max. payload 3 kN, date: 22.02.2004): loading in steps of 25 kg up to 300 kg, cf. section 3.2.1.1; right: zoomed region with dashed line showing the trend (mean value: 2046.57, standard error: ±0.468 (0.03%, calculated using GNUPLOT)

For the data acquisition with computer, an A/D-card was needed to convert voltage into digital units. The resolution of this A/D converter was 16 bit or $2^{15} = 65536$ values. Note, that the accuracy of these measurements depends on calibration and stability of the electronic signals.

Load cells

The maximum scatter in load cell signal was identified in the load cell calibration tests (load cell s.nr. 22 843, maximum load of 3 kN) as ±17 digital values. A linear regression satisfied the conversion of the digital values a with respect to force F

$$\text{Load cell s.nr. 22 843:}   F = 0.109a + 24.385 \quad (\text{in N}) \; . \qquad (3.12)$$

The scatter is equal to ±1.85 N, or refered to a minimal specimen area of approximately $A \approx 75 \times 75 \; \text{mm}^2 = 4417.9 \; \text{mm}^2$, equal to ±0.42 kPa. This load cell was used for triaxial tests with confining pressures up to 100 kPa. The maximum measured axial force was $F_{max}=1719.55$ N (test SAP1H). The average error[67] in the force measurement can therefore be assumed as

---

[67]Derivation see following remarks to the GDS device.

$\Delta F/F_{max} = \pm 0.11\%$, or $\Delta \sigma_a = \pm 0.08\%$. For tests with higher confining pressures, other load cells were necessary, see table 3.6.

Other electronic devices connected to the A/D-card were the two LVDTs, a pressure transducer (only for tests with low confining pressures) and the photoelectric relays of the U-shaped observation tube.

permissible pay load of the used load cell	conversion equation (N)	errors in axial	
		force ΔF	stress $\Delta \sigma_a$
3 kN (s.nr. 22 843) for tests with $\sigma_c \le 100$ kPa	$F=0.109a+24.385$	$\pm 0.11\%$	$\pm 0.08\%$
25 kN (s.nr. 807 156) for the other tests[a]	$F=0.870a+446.6$	$\pm 0.19\%$	0.15%

[a] the 'worst case' is a test with low axial force due to low confining pressure, such as SAP2F (max.F=7730.95 N → max.σ_a=7730.95 N/4417.9 mm^2+500 kPa=2249.9 kPa)

Tab. 3.6: Scatter in load cell measurements

LVDTs

The LVDT measurement range was 10 mm. The calibration of the LVDTs lead to the linear conversion[68] of digital units a to deformation s:

$$\text{LVDT 1, s.nr. 6297:} \quad s \text{ (in mm)} \quad = \quad -0.00019a \;, \quad (3.13)$$
$$\text{LVDT 2, s.nr. 6382:} \quad s \text{ (in mm)} \quad = \quad -0.00019a \;. \quad (3.14)$$

The scatter in the acquired data was maximal ± 30 digital values, see also Fig. 3.28 (page 100) which is equal to approximately ± 0.0057 mm or $\Delta \varepsilon_a = \pm 0.011\%$ (for an initial distance of the LVDT mounting parts of 50 mm).

Photoelectric relays

The accuracy of the U-shaped observation tube was already estimated in section 3.1.3. The scatter in volume is maximal ± 15 mm^3 (corresponding to $\Delta \varepsilon_v = \pm 0.01\%$, assuming a specimen volume of 330 cm^3 and an average void ratio of e=0.5).

---

[68]With the LVDTs a calibration sheet was delivered. The linearity was certified with an accuracy of $\le 0.10\%$.

GDS controller

The GDS device for the control and acquisation of the confining pressure had a measurement resolution of ± 1 kPa. It was connected directly to the serial port of the computer, the A/D converter was not necessary. This scatter can be assumed to be of minor importance for tests with high confining pressures, such as 100 kPa or more. For example[69] for tests with 100 kPa confining pressure the scatter leads to an error in the cell pressure of maximal $\pm 1\%$, or applied on the axial stress (of exemplarily test SAP1H with σ_a=1719.55 N/4417.9 mm^2+100 kPa=506.30 kPa) to a maximal error of $\pm 0.2\%$. For the tests with low confining pressures (< 100 kPa) the cell pressure was additionally measured with a pressure transducer.

Summary

The overall average error due to scatter in measurement devices cannot be defined for one specific evaluated dimension, such as deviatoric stress q. The overall error due to scatter in measurement devices is assumed to be $<1\%$, see table 3.7 (σ_c). Note that to the author's opinion the error caused by scatter in measurement devices is of minor importance. The variability of measured soil properties caused by specimen preparation has more influence on the scatter in results.

---

[69]For reasons of simplicity the 'worst combination' of force and area — which has to lead to the maximum error — was chosen, although this combination did not appear in the real test, where the area is larger due to lateral expansion of the specimen and subsequent lower axial stress. This is only an illustrative example, to show the influence of the errors.

Measurement	Calculated value	Maximal deviation in calculation	Remarks
axial deform. Δs_a	axial strain ε_a	$\pm 0.011\%$	
	specimen area A_m	$\pm 10^{-3}\%$	
	axial stress σ_a	$\pm 0.01\%$	Membrane corr., see chap. 3.2.1.2.
volume change ΔV	volumetric strain ε_v	$\pm 0.01\%$	
	specimen area A_m	$\pm 10^{-3}\%$	
axial force F_a	axial stress σ_a	$\pm 0.19\%$	
	axial strain ε_a	$\pm 0.01\%$	Compliance, see chap. 3.2.1.6.
confining pressure σ_c	axial stress σ_a	$\pm 0.2\%$	
	lateral stress σ_r	max.$\pm 1\%$	With σ_c=100 kPa, else smaller.

Tab. 3.7: Effect of the scatter on the results of the triaxial tests

3.3.4 Soil variability

Soil and soil properties vary not only in layers, but also in apparently homogeneous deposits. For the illustration of soil variability some coefficients of variation for selected soil properties after Baecher and Christian[70,71] can be found in table 3.8.

Test	Reported COV (%)	Standard
Air voids	16-30	20
Angle of friction (sand)	5-15	10
Cohesion (undrained, sands)	25-30	30
Density (apparent or true)	1-10	3
Elastic modulus	2-42	30

Tab. 3.8: Coefficients of variation for soil properties, taken from Baecher and Christian[70]

In this study the main interest is on the bulk properties and on the strength of laboratory specimen (or models) and field prototypes.

---

[70] Baecher, G.B., Christian, J.T.: Reliability and Statistics in Geotechnical Engineering, John Wiley & Sons Ltd., Chichester, 2003

[71] For reasons of simplicity here only the few soil properties which were basic for this study. Further informations and more properties can be found in the publication of Baecher and Christian.

Strength parameters

The influence of measurements and devices on the results of the triaxial tests was shown in detail in section 3.2.

Bulk properties

The unit weight γ_s of a soil can be measured very accurate with conventional laboratory equipment. The error is reasonably small. The dry density γ_d varies in field over a large range, see table 3.8. Therefore, it is utterly impossible to assign *one* density to a soil mass. Also in laboratory models or reconstituted specimen the density can vary, i.e. see Fig. 3.37 on page 106.

For the modelling of a prototype situation one has to know the density in-situ. Methods for the determination of density in-situ are i.e. substitutional methods (water or plaster substitution), or extraction methods. The latter ones are problematic for cohesionless soils.

In-situ determination of relative density: Direct methods for density determination are extraction of undisturbed samples, replacement methods, such as water, sand or plaster replacements, or sampling after freezing of the soil. Indirect measures are correlations with e.g. penetration resistance, wave velocities, attenuation of radiation (nuclear methods), or bearing plate tests. Indirect methods give only a rough estimation of the relative density.

The relative density of a huge soil mass in-situ can be estimated only from random samples, e.g. using the standard penetration test, where a probe is dynamically driven into the soil. The compactness of the soil and the standard penetration index are correlated.This correlation is often used in engineering practice.

After Tavenas and LaRochelle[72] it is virtually impossible to assign a density to the standard penetration index N. In Fig. is shows the scatter in the correlation between N and D_r, where the relative density of this soil could be between $D_r=56$ and 87% at 25 blow counts N due to the immanent error in N, thus in this case the soil could be medium dense ($35\%< D_r <65\%$) or dense ($85\%< D_r \leq 100\%$). With such large scatter the deduced relative density becomes useless.

Laboratory methods for the determination of relative density: For the definition of relative density in laboratory exist various experimental methods to

---

[72]Tavenas, F., LaRochelle, P.: Accuracy of relative density measurements, *Géotechnique*, 4 (**22**) 1972, pp. 549-562

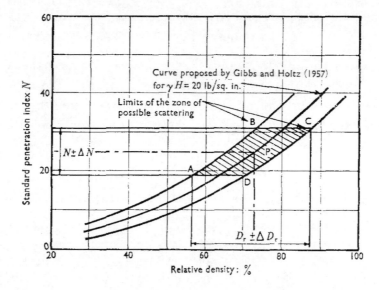

Fig. 3.51: Scattering in the correlations between N and D_r, taken from Tavenas and LaRochelle[72]

determine the limiting densities. The densest compaction is usually provided by one of the following methods[73] (a) using vibrating tables, or (b) using a fork for vibration, the loosest one by filling a cylindrical mould with the help of a funnel or a shovel with soil.

However, Tavenas and LaRochelle[72] found an error of about ±6% in the determination of D_r in best laboratory conditions and ±13% in field conditions.

Sladen and Handford[74] report on errors in the triaxial testing of very loose sands. Usually, these specimens can be prepared from unsaturated sand due to an apparent cohesion. The density of the specimen is determined from its weight and from the dimensions *before* the final assembly of the triaxial cell. Due to the saturation the specimens densify. This volume change is usually not measureable while assembling the cell, but influences all computations of density during all test stages. This error might also apply to laboratory model tests, if with saturated loose soils is worked and the dimensions are only measured prior the test. It is therefore highly recommended to monitor the dimensions during saturation of such specimens. Sladen and Handford found an error up to Δe=0.15 overestimation (which is in the case of Syncrude

---

[73]e.g. German Standard DIN 18 126, US Standard ASTM-D4253-00

[74]Sladen, J.A., Handford, G.: A potential systematic error in laboratory testing of very loose sands, *Canadian Geotechnical Journal*, 3 (**24**) 1987, pp. 462-466

tailings sand an error of 20%!). With Soiltron the loose packing is obtained with light additives, and therefore stable.

The in-situ to such an extent poorly determinable, but nevertheless the soil behaviour mainly characterising parameter: the density, is to the author's opinion the main source of misinterpretations of measurements in tests in-situ or measurements from model test which were prepared according to the *wrongly* assessed in-situ density.

3.3.5   Verification of area correction assumptions by numerical study

A numerical study was performed in order to assess the appropriateness of the assumptions taken for the calculation of mid-plane area in the evaluation of the triaxial tests (section 3.2.1.3).

To realistically describe the mechanical behaviour of sand realistic, the hypoplastic constitutive law was used for a finite element analysis with ABAQUS. The analysis was subdivided into the following load steps according to the usual triaxial test procedure:

1. Initial test conditions:

    — Self-weight of the specimen
    — Vacuum pressure (5 kPa) to stabilize the specimen
    — Axial load ba top plate weight

2. Isotropic consolidation
3. Shearing up to 25% axial strain

The following effects of different types of end restrains between specimen surface and end plates were examined:

1. SC: smooth contact (perfect lubrication)
2. FC: frictional contact, soil-platen friction angles of 2°, 5° and 10°

The triaxial specimen was discretized into 100 8-node biquadratic axisymmetric quadrilateral elements. The specimen's top surface is in contact with a rigid surface which can only move vertically, while the bottom is in contact to a fixed rigid surface. The right and upper surfaces are subjected to a circumferential pressure, in the first calculation step a pre-consolidation pressure of 5 kPa which is increased in the second calculation step up to the consolidation pressure of 500 kPa. The compression (third calculation step) is applied by lowering the upper rigid surface. The final value of axial strain of the compression is 25%. The soil was simulated with the properties as summarized in table 2.5 on page 69.

The FE-results are compared with the real test SAP2E, as evaluated in section 4.1. In the figures 3.52 the shape of the triaxial specimen SAP2E is sketched (i) as raw data, measured from local and global deformation measurements using LVDTs and on the cell mounted displacement transducers (RD), (ii) considering local LVDT measurements up to onset of inhomogeneous deformation and global measurements afterwards assuming the decay of bedding error and barreling (AS), as described in section 3.2.1.3, and (iii) from finite element calculations (SC and FC). For the purpose of comparison the surface shapes at same magnitude of axial strains are plotted. Due to bedding error, the global axial strain in the raw data diagram is higher than the local and is therefore annotated in diagram RD.

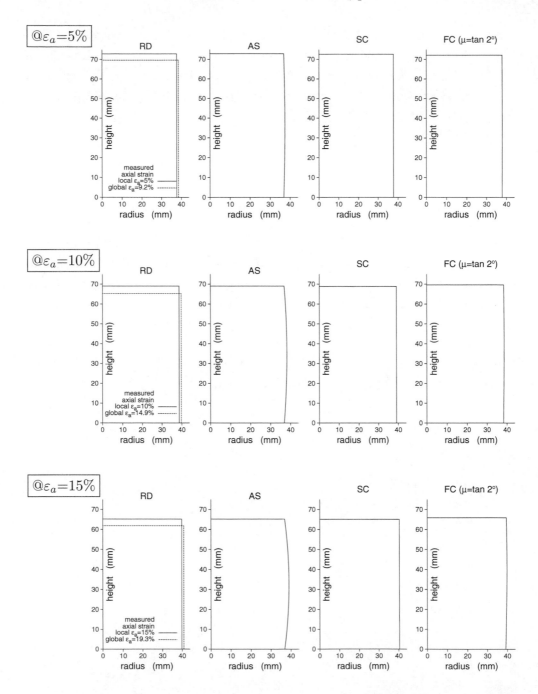

Fig. 3.52: Calculated deformation of specimen SAP2E at ε_a=5%, 10% and 15%; raw data (RD), assumed shape for evaluation of tests (AS), FE – smooth contact (SC), FE – frictional contact (FC) with friction $\mu = \tan 2°$ (exemplarily)

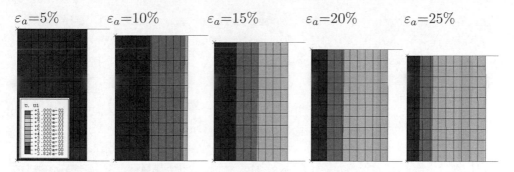

Fig. 3.53: Spatial distribution of horizontal deformation from FE – smooth contact (SC) simulation

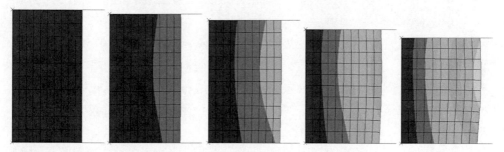

Fig. 3.54: Spatial distribution of horizontal deformation from FE – frictional contact (FC); $\mu = 2°$

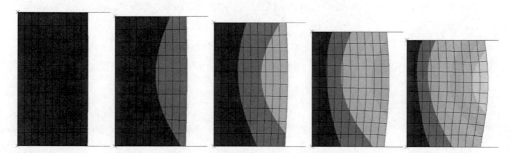

Fig. 3.55: Spatial distribution of horizontal deformation from FE – frictional contact (FC); $\mu = 5°$

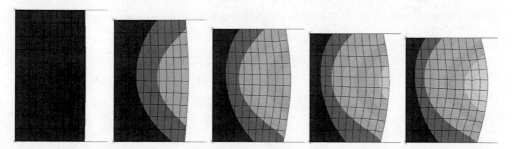

Fig. 3.56: Spatial distribution of horizontal deformation from FE – frictional contact (FC); $\mu = 10°$

The end plate friction is hard to assess, since it is viscous due to lubrication. For the present simulations, dry friction with various friction coefficients has been assumed, see Fig. 3.54-3.56.

The figures 3.57-3.60 show the evolution of axial stress at the horizontal midplane of the specimen, as evaluated from test data and from the FE-simulations. The stress distributions are in the cases RD and AS assumed as homogeneous over the midplane and for the FE-simulation of smooth contact (SC) evaluated as homogeneous over the midplane. The main advantage in using both deformation measurement (model AS) is that test data can be processed for large deformations. Local strain gauges are limited in measurement range and after onset of barreling of the specimen not reliable. If one consideres in later test stages the global strain gauges, the strain ranges are much more increased.

The simulations with frictional contacts (FC) show that the axial stress decreases from the center to the periphery of the specimen, as can be seen from figures 3.59-3.60.

From this investigation it can be concluded that if homogeneous deformation of the specimen is considered, the axial stress is overestimated (see Fig. 3.58) compared to the assumed deformation shape, which is additionally plotted in this figure. The volume change prediction using model SC is not realistically. The axial stress, calculated with the assumed deformation shape (AS) coincides quite well with the numerical results that consider friction on the end-plates.

A small end plate friction leads to barreling of the specimens which was mostly encountered in the test series, which is taken into account for this comparison, as can be seen from Fig. 3.61. Test SAP2E exhibits inhomogeneous deformation. After testing the size of the specimen was measured and it was photographed. It must be remarked that the measurement is difficult due to the local bulging of the membrane which is too long compared to the specimen after the shortening of the specimen. Another source of error is that due to the release of cell pressure the specimen slightly deforms. From the photograph Fig. 3.61a one can see the barreling of the specimen.

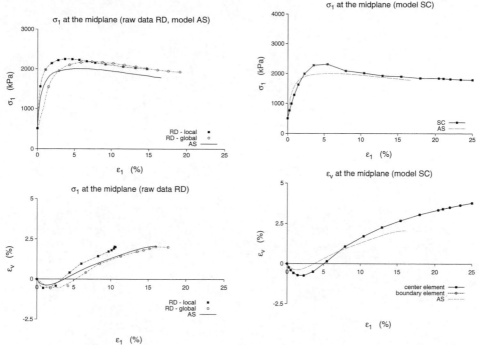

Fig. 3.57: Calculated axial stress σ_1 and volumetric strain ε_v from raw data (RD) and area assumptions (AS)

Fig. 3.58: FE-simulation of axial stress σ_1 and volumetric strain ε_v (SC: smooth contacts, AS: assumed shape in evaluation of the triaxial tests)

Additionally, Fig. 3.61b shows the measured shape of specimen SAP2E after the tests and the assumed specimen geometry for the calculations. The assumption coincides quite well with the measurement after dismantling of the triaxial cell. The axial deformation measurement using LVDT were in ths case of test SAP2E considered up to an axial strain $\varepsilon_a=2.5\%$ and subsequently from the global deformation measurement using the strain gauge mounted outside the cell, Fig. 3.61c.

The results of the FE-simulations obtained with a small friction at the end platens agree quite well with the results taking into account the assumptions of section 3.2.1.3. The specimen deformation of test SAP2E is typical for the observations of the triaxial tests. Therefore, for the evaluation of all tests the homogeneous deformation with subsequent barrelling was assumed.

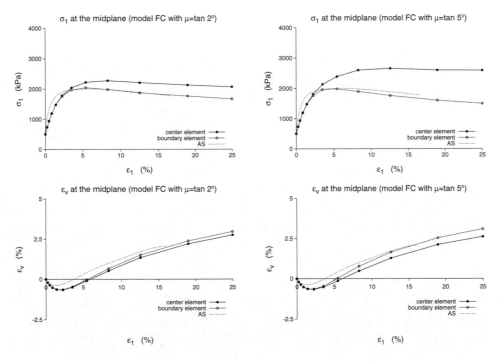

Fig. 3.59: FE-simulation of axial stress σ_1 and volumetric strain ε_v at the center and periphery elements (FC: frictional contact for the contact frictions $\mu = \tan 2°$ and $\tan 5°$, AS: assumed shape in evaluation of the triaxial tests)

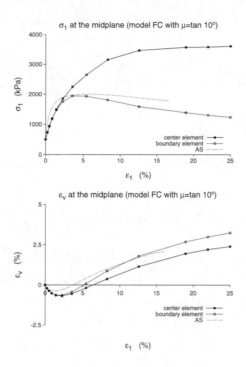

Fig. 3.60: FE-simulation of axial stress σ_1 and volumetric strain ε_v at the center and periphery elements (FC: frictional contact for the contact frictions $\mu = \tan 10°$, AS: assumed shape in evaluation of the triaxial tests)

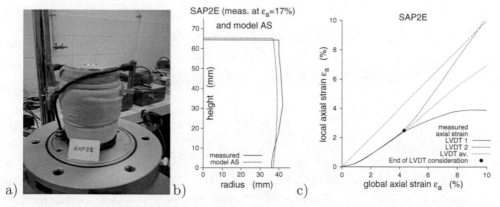

Fig. 3.61: Specimen SAP2E after the triaxial test, comparison of measured and assumed specimen shapes in the calculations, consideration of LVDT-measurements for model AS

Chapter 4

Test results and implications for Soiltron

4.1 Overview of tests and results

The test conditions and basics for the evaluation of the test results are detailed described in Chapter 3. The volume change was measured using a GDS standard volume/pressure controller and the U-shaped observation tube, as described in section 3.1.3. The tests were evaluated taking into account the membrane penetration and stiffness, influence of compliances of the loading system, and the area corrections, see section 3.2. The sand specimens were prepared in layers as described in section 3.1.5.

4.1.1 Triaxial tests with Ottendorf-Okrilla sand

The mechanical behaviour of Ottendorf-Okrilla sand 0.1-2 mm was investigated in:

- SAP$1x$-$5x$: constant σ_r triaxial tests
- SASP: $\dot{\varepsilon}$-jump, conventional triaxial tests with change of strain rate / piston velocity
- SACPx: conventional triaxial tests with unloading and reloading (one-way cyclic tests)
- K_0-compression, see section 3.2.1.7

In table 4.1 are summarised the initial conditions and results of the triaxial tests with dry Ottendorf-Okrilla sand (series SAP). The abbreviation means: **SA**nd which is tested in **P**rague. Preliminary tests were conducted in Innsbruck in 'simple' conventional triaxial devices (SAI). The results of these tests are not presented in the thesis.

Test No.	initial values		cell pres-sure	after conso-lidation		peak values			critical friction angle
	void ratio	dry density		void ratio	dry density	friction angle	strain	angle of dila-tancy	
	e_0	γ_d	σ_r	e_c	$\gamma_{d,c}$	φ'_p	$\varepsilon_{a,p}$	ψ'_p	φ'_c
	(−)	(g/cm^3)	(kPa)	(−)	(g/cm^3)	(°)	(%)	(°)	(°)
SAP1B	0.529	1.723	100	0.507	1.749	39.6	3.6	12.1	32.2
SAP1C	0.514	1.741	100	0.492	1.766	39.4	4.0	12.2	
SAP1D	0.525	1.726	100	0.514	1.741	40.2	3.8	13.5	*
SAP1G	0.522	1.732	100	0.503	1.753	39.4	5.0	10.6	
SAP1H	0.521	1.733	100	0.505	1.751	40.2	5.6	12.5	
SAP2C	0.521	1.733	500	0.473	1.788	37.1	5.8	8.5	33.5
SAP2E	0.538	1.714	500	0.486	1.773	36.9	6.2	7.6	*
SAP2F	0.497	1.761	500	0.457	1.808	38.8	5.3	10.4	34.8
SAP2G	0.527	1.726	500	0.487	1.772	36.8	5.3	7.1	
SAP3B	0.537	1.714	1000	0.475	1.772	34.6	8.0	4.1	32.3
SAP3D	0.518	1.736	1000	0.448	1.820	36.5	6.9	2.5	32.1
SAP3G	0.483	1.777	1000	0.430	1.843	36.6	5.1	5.1	*
SAP3H	0.475	1.786	1000	0.427	1.847	36.9	6.8	7.1	
SAP3I	0.469	1.794	1000	0.425	1.849	36.2	7.2	6.5	
SAP3J	0.509	1.746	1000	0.471	1.791	35.4	8.9	4.9	
SAP4A	0.494	1.764	2000	0.436	1.835	33.5	8.7	1.7	*
SAP4C	0.530	1.722	2000	0.456	1.810	33.5	8.7	3.3	32.1
SAP5B	0.540	1.711	1500	0.489	1.770	33.5	8.7	1.9	32.7
SAP5E	0.493	1.765	1500	0.454	1.812	35.0	7.2	1.1	32.8
SAP5F	0.496	1.761	1500	0.449	1.812	35.0	8.4	3.8	33.9
SASP	0.479	1.782	1000	0.431	1.841	-	-	-	-
SACP1	0.530	1.722	1000	0.472	1.790	-	-	-	-
SACP2	0.513	1.742	1000	0.455	1.811	-	-	-	-

*... The critical state was not reached, see Fig. 4.3.

Tab. 4.1: Initial conditions and results of successful tests with Ottendorf-Okrilla sand (series SAP)

| Max. void ratio | e_{max} | 0.75 |
| Min. void ratio | e_{min} | 0.42 |

Tab. 4.2: Minimal and maximal void ratios of Ottendorf-Okrilla sand (according to German standard DIN 18 126)

The triaxial specimens were isotropically consolidated. Fig. 4.1 shows the compression lines (void ratio e vs. mean pressure p') of these tests. Further in this diagram the change of e during the shearing is shown. The isotropic consolidation is marked by dashed lines, and the shearing by continuous lines. The points at the end of some tests mark the ends of tests where it was, according to the test results, assumed that a critical state was reached ($\Delta\varepsilon_v \approx 0$) in partitions of the specimens. The three lines (e_i-, e_c- and e_d-lines) characterise the densest (e_d) and the loosest possible (e_i, or isotropic normal compression line) states of the soil. There are only few tests which exhibit higher void ratios than e_c. They can be attributed to scatter in measurements (even the initial density is difficult to assess).

Remote measurement of deformations using evaluation of photographies with the PIV method, see section 3.1.4, and also numerical simulations, see section 3.3.5, showed nonhomogeneous deformations of the specimens. The critical state is not uniformly reached in triaxial tests. Anyway, there is a good coincidence between the marked points and the critical state line which was determined in the application of equ. (2.4), determined with the oedometric compression tests of section 2.3, and applied with $e_{c0} \approx e_{min}$.

Fig. 4.2 shows deviatoric stress-strain curves and the volumetric strains of the triaxial tests with Ottendorf-Okrilla sand. The axial strain is defined as $\varepsilon_a = \Delta h/h_0$, with h_0 being measured with the LVDTs and the outside the cell mounted dial gauge. The transition from 'local' (LVDT) to 'global' (outer dial gauge) measured strains was defined for every test differently according to the procedure which is described in section 3.2.1.3. The tests were stopped in the case of errors, such as shifting of specimens or leakage of the membrane.

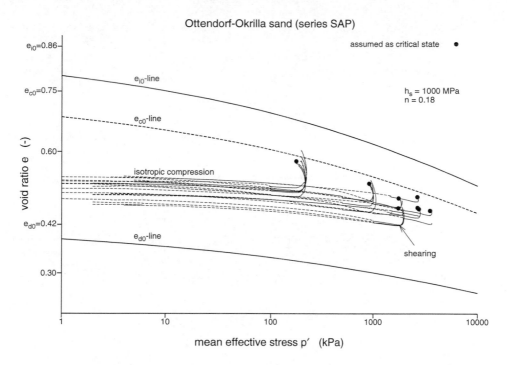

Fig. 4.1: Compression lines of isotropic consolidations of the tests SAP. Also plotted: upper and lower boundaries of possible void ratios e_i, e_d for: $h_s = 1000$ MPa, $n = 0.19$) according to equ. (2.4), section 2.3

Another depiction of the test results is shown in Fig. 4.3, where the friction angles and void ratios are plotted. The mobilized friction angle in Fig. 4.3 is defined as

$$\varphi = \sin^{-1}\left(\frac{\sigma_1 - \sigma_2}{\sigma_1 + \sigma_2}\right) \quad . \tag{4.1}$$

As can be seen from this figure, the peak friction angle decreases with increasing confining pressure from approximately $40°$ to $33°$ due to the decreasing relative density r_e, see Appendix A.1. For tests with the lowest investigated confining pressures (100 kPa) a pronounced volume increase (dilatancy) was found. The tendency of dilatancy decreases with increasing confining pressure, which can be explained with Fig. 4.1 where after the consolidation the state of the specimens was already close to the critical state line.

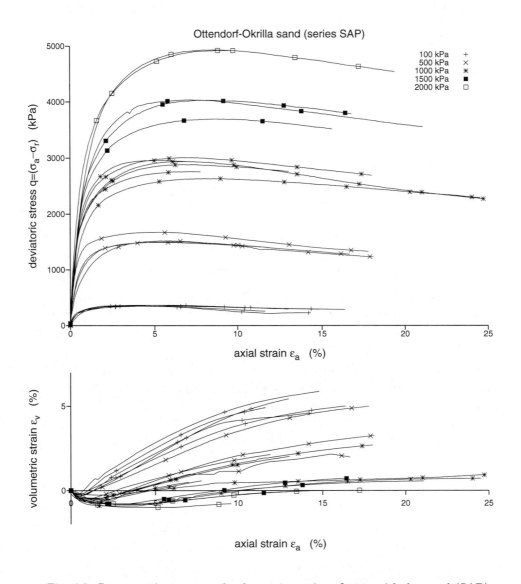

Fig. 4.2: Stress-strain curves and volumetric strains of tests with dry sand (SAP)

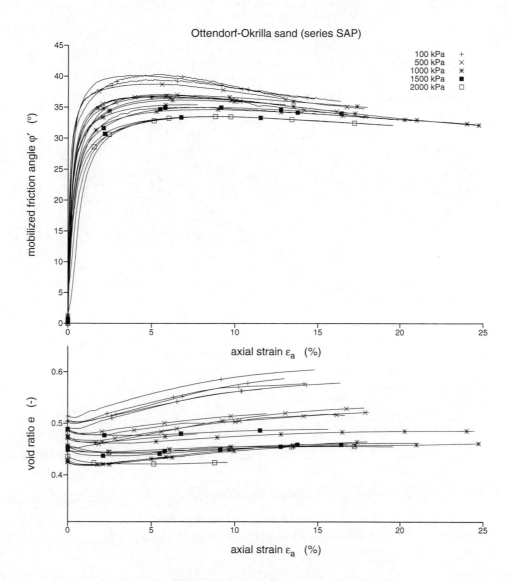

Fig. 4.3: SAPx: friction angle φ_{mob} and void ratio e

The mechanical behaviour of soil can be described by the quantities peak friction angle φ'_p, axial strain $\varepsilon_{a,p}$ at φ'_p, dilatancy angle ψ'_p at the point φ'_p and the critical friction angle φ'_c. According to Schanz and Vermeer[1] ψ was calculated as

$$\sin \psi = -\frac{\dot{\varepsilon}_v / \dot{\varepsilon}_1}{2 - \dot{\varepsilon}_v / \dot{\varepsilon}_1} \qquad (4.2)$$

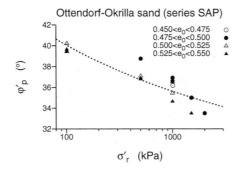

Fig. 4.4: Peak friction angle φ'_p vs. cell pressure σ'_3 in dependence of initial void ratio e_0

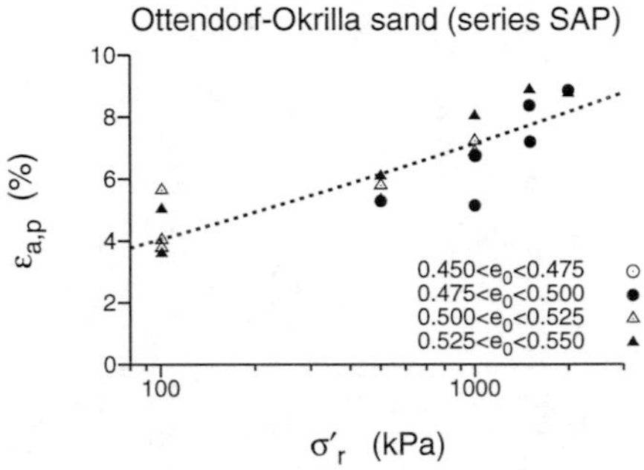

Fig. 4.5: Peak axial strain $\varepsilon'_{a,p}$ vs. cell pressure σ'_3 in dependence of initial void ratio e_0

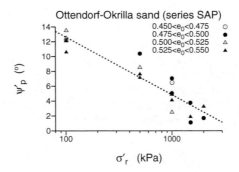

Fig. 4.6: Peak dilatancy angle ψ'_p vs. cell pressure σ'_3 in dependence of initial void ratio e_0

In Fig. 4.4 is shown the stress-dependence of peak friction angle. Due to the fact that the triaxial tests were done on reconstituted specimens — the initial density varied — no unique relation between pressure and peak friction angle can be found. For this reason the specimens with comparable void ratio

[1]Schanz, T., Vermeer, P.A.: Angles of friction and dilatancy of sand, *Géotechnique*, 1 (**46**) 1996, pp. 145-151

are marked with the same symbols. The line in the diagram was added to point out the general trend of the evolution of friction angle with increasing confining pressure. The same applies to Figures 4.5 and 4.6, where the strain $\varepsilon_{a,p}$ and the dilatancy angle ψ_p' are plotted versus confining pressure σ_r'. In section 4.2 will be presented normalizations which take pressure- and density-dependence of the test results into acount.

4.1.1.1 Influence of strain rate (SASP)

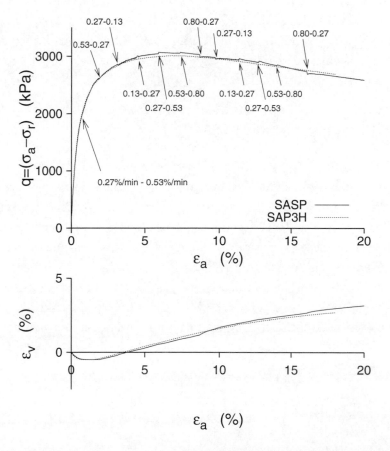

Fig. 4.7: Triaxial tests with similar initial void ratios e_0 and varying strain rate. Dotted line: test with constant strain rate (0.27%/min), continuous line: test with $\dot{\varepsilon}$-jumps. Values above the lines correspond to decreased strain rate and values below the line correspond to increased strain rate.

The influence of strain rate on the results in the triaxial test was investigated with a test where the strain rate during the test was increased and decreased, see Fig. 4.7. The initial velocity for this test was 0.2 mm/min (0.27%/min). The values in the diagram represent the $\dot{\varepsilon}$-jumps denoted in % of sample height per minute.

Note that changes of strain rate had to be limited to relatively low values to assure a precise measurement. Due to the relatively long reaction time of the GDS controllers for the measurement/control of pore-air and cell-water, relatively small jumps had to be chosen.

In addition, the result of a standard triaxial test with the same initial void ratio is shown in Fig. 4.7. As can be seen, no pronounced difference between the static loaded test (SAP3H) and the test with $\dot{\varepsilon}$-jumps (SASP) is visible. Thus, the in velocity range 0.1-0.6 mm/min (0.13-0.80%/min) no substantial effects on the mechanical behaviour are observed. The velocity 0.2 mm/min (0.27%/min) was chosen for all triaxial tests.

4.1.1.2   Unloading/reloading (SACPx)

Two tests with unloading/reloading cycles were conducted. Both experiments show quite similar behaviour, because both specimens were prepared with nearly the same initial density. We can see here that repeatable tests are possible, despite the use of the tamping method for the specimen preparation, see section 3.1.5. These tests were not done for a special reasons in this thesis, but were used for the check of the constitutive parameters of section 2.3. With the v. Wolffersdorf model of hypoplasticity the laboratory test cannot be perfectly reproduced, see Fig. 4.9. A better agreement could be obtained with a hypoplastic equation taking into account intergranular strain.

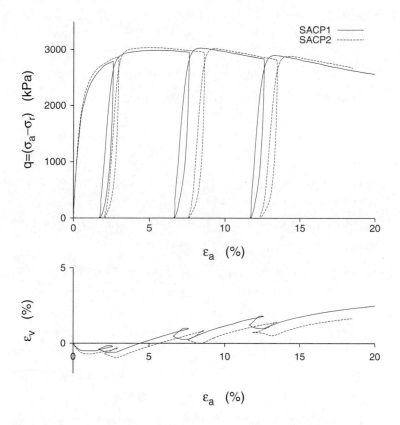

Fig. 4.8: SACPx: Triaxial tests with unloading and reloading

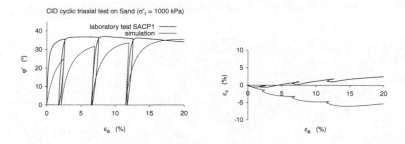

Fig. 4.9: Numerical simulation of tests SACP

4.1.2   Triaxial tests with Soiltron 1

The mechanical behaviour of Soiltron 1 (see section 2.2.2, page 53) was investigated in:

- SO1P1x-5x, SO1xPx: constant σ_r triaxial tests
- SO1P6A and SO1P7A: constant σ_a triaxial tests
- SO1P6B and SO1P7B: constant p' triaxial tests

The initial conditions and test results are summarised in table 4.3. Three different ratios of mixtures were investigated, see section 2.2.2. The mixtures are abbreviated with SO1Px which stands for **SO**iltron **1** (with polystyrene) tested in **P**rague. SOILP means with lower content of additives and SO1HP are the specimens with higher content of additives.

All specimens were isotropically consolidated. Fig. 4.10 shows the compression lines for the consolidation (dashed lines) and during the shearing stage (continuous lines). The points at the end of some compression lines are assumed as the critical state of these triaxial tests. In comparison to the compression of 'pure' Ottendorf-Okrilla sand (Fig. 4.1) is the noticeable softer response of the material visible. This is manifested also in the parameters h_s and n of equ. (2.4).

The range of compactness (table 4.4) is due to the additives different from that of the untreated sand. The relative densities of the Soiltron 1 specimens is close to the one of tests on sand. A complete list of pressure dependent densities r_e can be found in Appendix A.1.

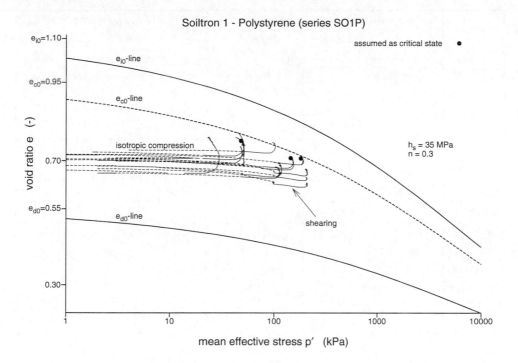

Fig. 4.10: Compression lines of isotropic consolidations of the tests SAP. Also plotted: upper and lower boundaries of possible void ratios e_i, e_d for: $h_s = 30$ MPa, $n = 0.33$) according to equ. (2.4), section 2.3

Figures 4.11 and 4.12 show the test results of the triaxial test in two different representations – deviatoric stress or mobilized friction angle vs. axial strain, and volumetric strain or void ratio vs. axial strain respectively.

In comparison to the triaxial tests on Ottendorf-Okrilla sand at 100 kPa confining pressure (SAP1) the dilatancy for Soiltron 1 is much lower. With decreasing cell pressure the dilatancy is increasing.

Test No.	initial values		cell pres- sure	after conso- lidation		peak values			critical friction angle*
	void ratio	dry density		void ratio	dry density	friction angle	strain	angle of dila- tancy	
	e_0	γ_d	σ_r	e_c	$\gamma_{d,c}$	φ'_p	$\varepsilon_{a,p}$	ψ'_p	φ'_c
	$(-)$	(g/cm^3)	(kPa)	$(-)$	(g/cm^3)	$(°)$	$(\%)$	$(°)$	$(°)$
Soiltron 1L (0.05 ml PS / 1 g sand)									
SO1LP1A	0.608	1.639	100	0.591	1.656	39.2	5.3	9.4	*
SO1LP2A	0.596	1.651	50	0.589	1.659	41.6	5.1	11.9	*
Soiltron 1 (0.1 ml PS / 1 g sand)									
SO1P1B	0.706	1.545	100	0.602	1.645	37.8	5.9	4.7	
SO1P1E	0.683	1.566	100	0.661	1.586	37.4	7.1	3.2	*
SO1P1F	0.674	1.574	100	0.649	1.598	37.7	6.3	5.8	
SO1P1G	0.721	1.531	100	0.625	1.622	35.3	5.9	5.3	32.6
SO1P2E	0.735	1.519	50	0.731	1.522	37.5	6.5	5.0	34.6
SO1P2F	0.685	1.564	50	0.656	1.591	40.2	4.7	4.1	*
SO1P2G	0.660	1.587	50	0.648	1.599	41.5	4.3	8.5	
SO1P3B	0.701	1.549	75	0.681	1.567	38.1	3.9	10.9	34.7
SO1P3C	0.700	1.550	75	0.676	1.572	38.0	4.0	8.3	
SO1P4A	0.710	1.541	20	0.695	1.554	43.1	3.7	13.1	
SO1P4C	0.719	1.533	20	0.708	1.543	44.0	4.5	8.9	*
SO1P4D	0.729	1.525	20	0.718	1.534	44.2	4.8	11.9	
SO1P5D	0.757	1.500	10	0.741	1.513	48.0	3.9	12.6	
SO1P5E	0.670	1.578	10	0.661	1.587	50.0	3.9	15.2	*
constant σ_a									
SO1P6A	0.692	1.558	$\sigma_a{=}100$	0.671	1.577	46.0	3.4	11.0	*
SO1P7A	0.742	1.512	$\sigma_a{=}50$	0.725	1.528	44.6	2.2	15.7	*
constant p'									
SO1P6B	0.735	1.519	$p'{=}100$	0.707	1.543	39.8	3.9	6.9	
SO1P7B	0.703	1.548	$p'{=}50$	0.690	1.559	43.3	3.7	15.9	
Soiltron 1H (0.15 ml PS / 1 g sand)									
SO1HP1A	0.783	1.478	100	0.759	1.498	35.2	7.8	1.5	34.3
SO1HP1B	0.814	1.453	100	0.784	1.477	35.2	9.1	0.0	34.0
SO1HP2A	0.792	1.471	50	0.780	1.481	38.1	6.5	2.4	

Tab. 4.3: Results of successful tests with Soiltron 1 (series SO1P)

| Max. void ratio | e_{max} | 1.00 |
| Min. void ratio | e_{min} | 0.54 |

Tab. 4.4: Minimal and maximal void ratios of Soiltron 1 (according to German standard DIN 18 126)

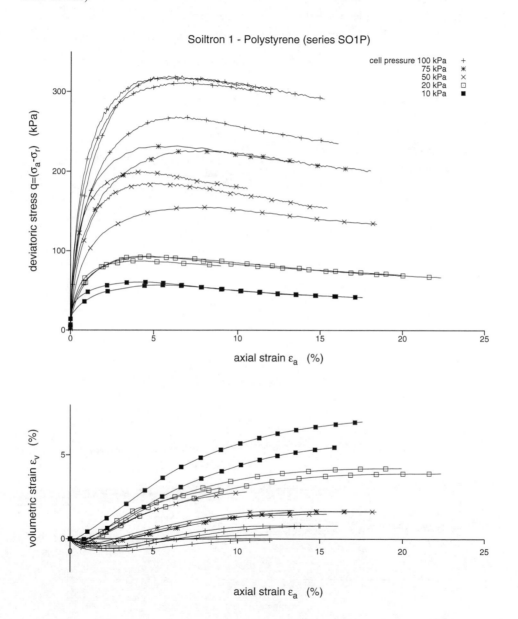

Fig. 4.11: Stress-strain curves and volumetric strains of tests with Soiltron 1 (SO1P)

The mobilized friction angle of Soiltron 1 is depicted in Fig. 4.12. From that diagram can be seen that for low confinement the peak value is much higher than for higher ones. The dilatancy for these tests is pronounced. The strength of the original material (pure sand) at the critical state is not visibly influenced by the additive polystyrene, the critical friction angle can be assumed as approximately 32° which is close to that of the test series SAP.

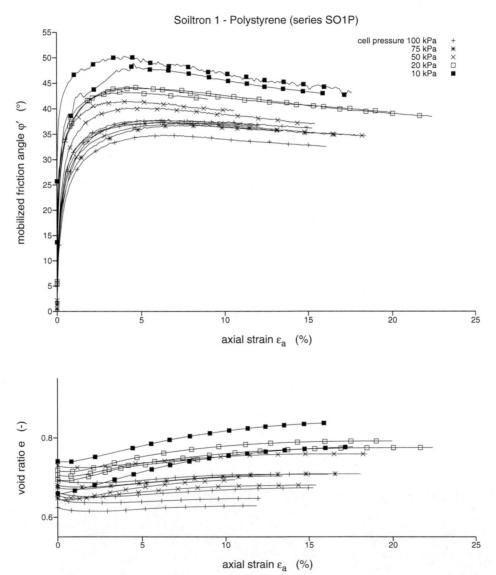

Fig. 4.12: SO1Px: mobilized friction angle φ_{mob} and void ratio e

The Figures 4.13-4.15 summarize the results of the triaxial tests on Soiltron 1. The material behaviour can be described with φ'_p, $\varepsilon_{a,p}$, ψ'_p. Noticeable is the stress and density dependence of these parameters. An appropriate normalization of these results which takes both into account was therefore seeked. This normalization is presented in section 4.2.

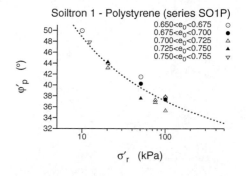
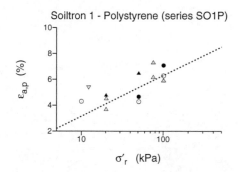

Fig. 4.13: Peak friction angle φ'_p vs. cell pressure σ'_3 in dependence of initial void ratio e_0

Fig. 4.14: Peak axial strain $\varepsilon_{a,p}$ vs. cell pressure σ'_3 in dependence of initial void ratio e_0

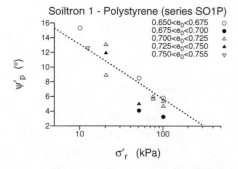

Fig. 4.15: Peak dilatancy angle ψ'_p vs. cell pressure σ'_3 in dependence of initial void ratio e_0

With conventional triaxial stress paths it is difficult to reach critical state at very low pressures, see Fig. 3.23, page 96. Therefore two further stress paths were investigated: constant σ_a and constant p'.

Constant σ_a triaxial tests **Constant p' triaxial tests**

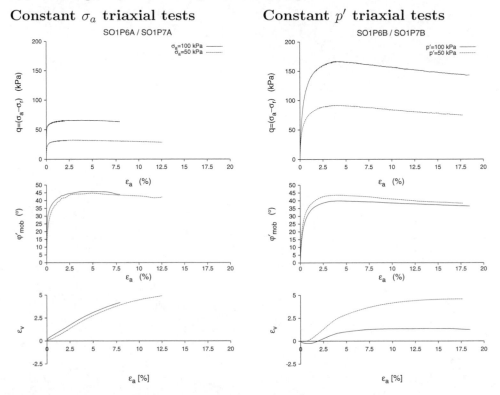

Fig. 4.16: Results of constant σ_a triaxial tests SO1P6A and SO1P7A: deviatoric stress q, mobilized friction angle φ' and volumetric strains ε_v vs. axial strain ε_a

Fig. 4.17: Results of constant p' triaxial tests SO1P6B and SO1P7B: deviatoric stress q, mobilized friction angle φ' and volumetric strains ε_v vs. axial strain ε_a

In Fig. 4.16 are shown the results of triaxial tests with constant σ_a. Test SO1P6A was started at 100 kPa cell pressure, SO1P7A at 50 kPa. With these tests a very low pressure level was reached at the peak. The measurements and subsequent control of the GDS cell water controller is very sensitive, a standard test would lead to a large scatter in the cell water pressure control with standard GDS controllers. The advantage of the constant σ_a-technique is that the cell water pressure is constantly lowered during the execution of the test, which can be more precisely controlled with the GDS controller. Also the constant p' triaxial tests, Fig. 4.17, lead to very low stresses at failure.

4.1.3 Triaxial tests with Soiltron 2

The mechanical behaviour of Soiltron 2 (see section 2.2.2, page 53) was investigated in:

- SO2P1x-5x, SO2xPx: constant σ_r triaxial tests
- SO2P7A: constant σ_a triaxial tests
- SO2P7B: constant p' triaxial tests

The tests were conducted similar to the aforementioned two test series.

The diagrams are provided in an analogous manner as for the test series with sand and Soiltron 1, and are therefore not described here in detail.

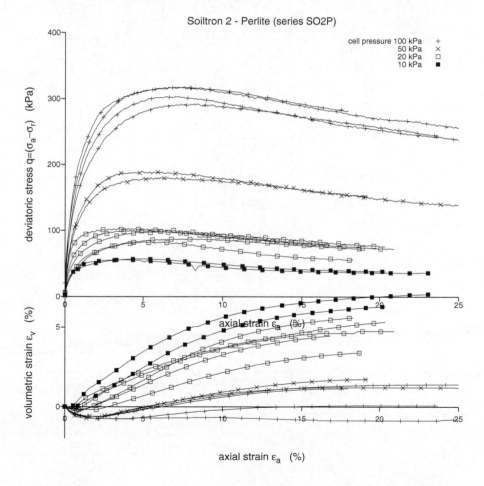

Fig. 4.18: Stress-strain curves and volumetric strains of tests with Soiltron 2 (SO2P)

Test No.	initial values		cell pres-sure	after conso-lidation		peak values			critical friction angle*
	void ratio	dry density	sure	void ratio	dry density	friction angle	strain	angle of dila-tancy	
	e_0	γ_d	σ_r	e_c	$\gamma_{d,c}$	φ'_p	$\varepsilon_{a,p}$	ψ'_p	φ'_c
	$(-)$	(g/cm^3)	(kPa)	$(-)$	(g/cm^3)	$(°)$	$(\%)$	$(°)$	$(°)$
Soiltron 2L (0.0067 g perlite / 1 g sand)									
SO2LP1A	0.629	1.618	100	0.609	1.638	38.9	6.0	7.5	*
SO2LP1B	0.609	1.638	100	0.588	1.660	39.8	5.9	9.3	35.3
SO2LP1C	0.600	1.648	100	0.586	1.662	39.4	6.5	7.2	*
SO2LP2A	0.597	1.650	50	0.586	1.661	41.3	4.3	11.2	*
Soiltron 2 (0.013 g perlite / 1 g sand)									
SO2P1A	0.732	1.522	100	0.673	1.575	36.9	6.6	0.0	32.5
SO2P1B	0.748	1.510	100	0.727	1.526	36.0	8.0	3.2	33.2
SO2P1C	0.727	1.526	100	0.698	1.552	37.7	7.4	4.9	33.5
SO2P1E	0.696	1.554	100	0.666	1.582	37.7	6.7	4.7	
SO2P2A	0.756	1.501	50	0.712	1.539	40.6	5.1	5.2	34.6
SO2P2B	0.755	1.502	50	0.725	1.528	39.8	6.3	5.8	
SO2P4A	0.721	1.531	20	0.710	1.541	44.8	6.4	10.1	
SO2P4B	0.698	1.552	20	0.688	1.561	45.4	3.5	7.6	
SO2P4C	0.718	1.534	20	0.707	1.544	44.8	4.4	9.7	*
SO2P4D	0.712	1.540	20	0.686	1.563	42.7	4.5	10.2	
SO2P4E	0.751	1.505	20	0.737	1.517	42.5	7.4	7.9	
SO2P5A	0.747	1.509	10	0.738	1.516	50.0	4.8	13.4	*
SO2P5B	0.651	1.596	10	0.643	1.604	48.1	3.7	15.1	
constant σ_a									
SO2P7A	0.753	1.503	50	0.737	1.517	46.8	3.5	11.7	
constant p'									
SO2P7B	0.690	1.560	50	0.675	1.574	43.3	5.3	10.2	
Soiltron 2H (0.020 g perlite / 1 g sand)									
SO2HP1A	0.850	1.425	100	0.817	1.450	35.5	11.5	0.0	34.4
SO2HP2A	0.819	1.449	50	0.796	1.468	39.1	8.2	3.2	

Tab. 4.5: Results of successful tests with Soiltron 2 (series SO2P)

Max. void ratio	$e_{\max}$	1.10
Min. void ratio	$e_{\min}$	0.63

Tab. 4.6: Minimal and maximal void ratios of Soiltron 2 (according to German standard DIN 18 126)

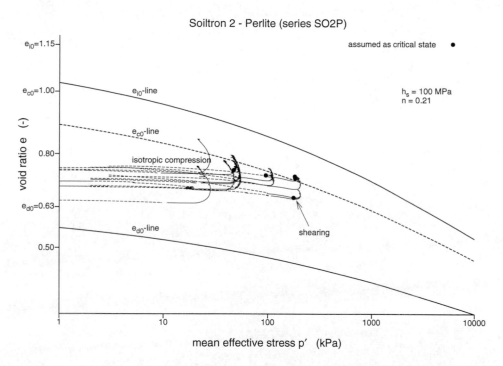

Fig. 4.19: Compression lines of isotropic consolidations of the tests SAP. Also plotted: upper and lower boundaries of possible void ratios e_i, e_d for: $h_s = 30$ MPa, $n = 0.21$) according to equ. (2.4), section 2.3

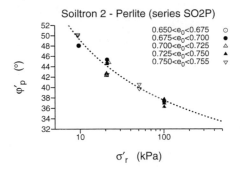

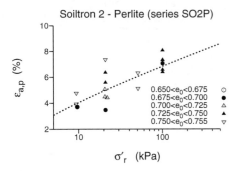

Fig. 4.20: Peak friction angle φ'_p vs. cell pressure σ'_3 in dependence of initial void ratio e_0

Fig. 4.21: Peak axial strain $\varepsilon'_{a,p}$ vs. cell pressure σ'_3 in dependence of initial void ratio e_0

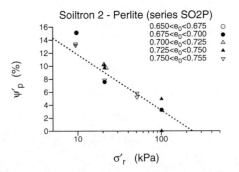

Fig. 4.22: Peak dilatancy angle ψ'_p vs. cell pressure σ'_3 in dependence of initial void ratio e_0

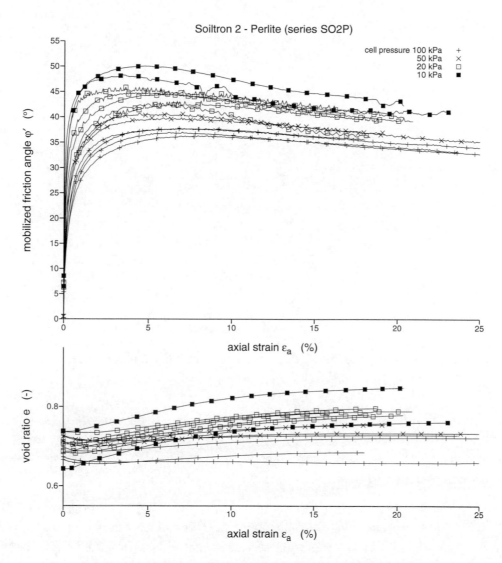

Fig. 4.23: SO2Px: mobilized friction angle φ_{mob} and void ratio e

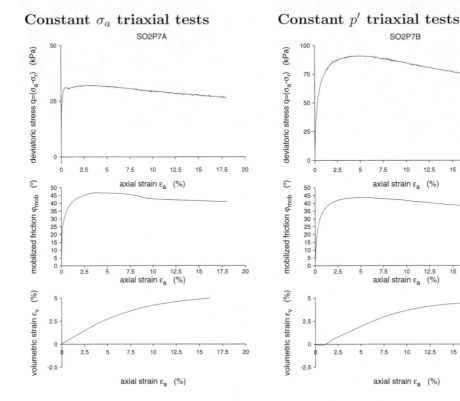

Constant σ_a triaxial tests   **Constant p' triaxial tests**

Fig. 4.24: Results of constant σ_a triaxial test SO2P7A: deviatoric stress q, mobilized friction angle φ' and volumetric strains ε_v vs. axial strain ε_a

Fig. 4.25: Results of constant p' triaxial test SO2P7B: deviatoric stress q, mobilized friction angle φ' and volumetric strains ε_v vs. axial strain ε_a

In Fig. 4.24 are shown the results of triaxial tests with constant σ_a. The test was started at 50 kPa cell pressure. Also the constant p' triaxial tests, Fig. 4.25, lead to very low stresses at failure. With both tests, the critical state was assumed to be reached, because ε_v is constant.

Due to the scatter in initial density of the specimen, no unique dependence between pressure and the soil behaviour characterising parameters φ'_p, $\varepsilon_{a,p}$ and ψ'_p was found. In the following section a method is presented that normalises the results with respect to the density and the pressure. With these normalisations relations can be developed that allow to calculate a required density for an intended mechanical behaviour at a given pressure state.

4.2   Appropriateness of Soiltron for 1g model tests

Soiltron can be used in small scale 1g tests if its mechanical properties are comparable to the mechanical properties of the prototype, which is here assumed as cohesionless soil. The main mechanical characteristics of such soils can be expressed by the following quantities:

- peak friction angle φ'_p
- peak strain $\varepsilon_{a,p}$
- dilatancy angle at peak ψ'_p
- φ'_c

The fact that these values (except φ'_c) are stress-dependent prohibits to use the prototype soil for a geometrically reduced model test, if the stresses are due to gravity. In the results of the triaxial tests, see tables 4.1-4.5, can be assumed that φ'_c is equal for sand and Soiltron.

The experimentally obtained relations φ'_p, ψ'_p, $\varepsilon_{a,p}$ vs. σ'_r (Figs. 4.4-4.6, 4.13-4.15 and 4.20-4.22) exhibit a large scatter, which is partly due to the fact that the initial void ratios e_0 of the several specimens are — for practical reasons — not completely coincident. It is therefore seeked to adjust the influence of e_0 by introducing an appropriate normalisation. The void ratio after consolidation e_c is chosen, because the specimens were consolidated to different stress states and are, therefore, densified. The values of φ'_p, ψ'_p and $\varepsilon_{a,p}$ are plotted versus appropriately chosen functions $f_i(\sigma'_r, e_c)$ rather than being plotted vs. σ'_r of the triaxial tests. These functions will enable to calculate an appropriate density of Soiltron for given model dimensions and simulated field conditions (prototype stress level and density).

These functions are basically of the form

$$f_i(\sigma'_r, e_c) = \alpha \cdot \left(\frac{\log \sigma}{e_c} \right)^{\beta} \quad , \tag{4.3}$$

α and β being constants and σ a dimensionless substitute of σ'_r. Note that α and β are not exponents for the hypoplastic constitutive equation. The functions were determined based on results of triaxial tests with confining pressure $\sigma'_r = 100$ kPa, see e.g. Fig. 4.26 which shows the peak friction angle φ'_p in dependence of the void ratio after consolidation e_c, where the test results and a regression line of these tests are plotted. Equ. (4.3) enables to plot the relations between φ'_p (or $\varepsilon_{a,p}$, ψ'_p) and e_c for various confining pressures (Fig. 4.27).

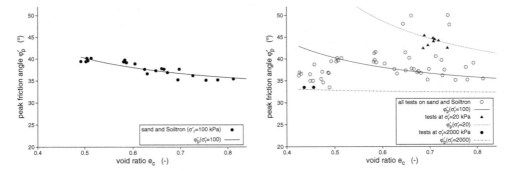

Fig. 4.26: Example for the derivation of equ. (4.3): Peak friction angles from triaxial tests on Ottendorf-Okrilla sand and Soiltron at confining pressure $\sigma'_r = 100$ kPa; regression line

Fig. 4.27: Example for the derivation of equ. (4.3): Results of all tests at various confining pressures; Lines showing extrapolated stress level/density dependence exemplarily for $\sigma'_r = 20$ kPa and $\sigma'_r = 2000$ kPa

These functions enable to calculate the required density of the model soil in the $1g$ small scale test to reproduce a prototype situation of known density and stress level.

4.2.1 Normalisation of φ'_p

The peak friction angle at a given confining pressure decreases nonlinearly with increasing void ratio (pyknotropy) and has the lower bound φ'_c (critical state). Furthermore, with increasing stress the peak friction angle decreases (barotropy). The following function is capable to model both effects as experimentally obtained for Ottendorf-Okrilla sand and Soiltron:

$$\sin\varphi'_p(\sigma'_r, e) = \sin\varphi'_c + \alpha_\varphi \cdot \left(1 + \log(\sigma_{ref}/\sigma'_r)\right) \cdot \left(\frac{\log\left(\sigma_{ref}/\sigma'_r\right)}{e}\right)^{\beta_\varphi} \quad , \quad (4.4)$$

with $\sigma_{ref} = 10000$ kPa being a reference stress. The parameters α_φ and β_φ can be calibrated using the void ratio vs. peak friction angle relation of triaxial tests with constant confining pressure over a wide range of void ratios, see line in Fig. 4.28. φ'_c was assumed as $32°$. The parameters for Ottendorf-Okrilla sand 0/2 and Soiltron 1 and 2 are[2]: $\alpha_\varphi = 4.2 \cdot 10^{-3}$, $\beta_\varphi = 1.600$.

---

[2]from regression analysis using GNUPLOT

The accuracy of equ. (4.4) can be seen in Fig. 4.30 where the measured φ'_p and the calculated ones are plotted, the deviation is less than 2° for most of the test results. Fig. 4.29 shows the dependence of peak friction angle on the confining pressure, the line shows exemplarily equ. (4.4) for $e_c = 0.5$. The tests SAP1C and SAP3B happened to have e_c-values close to 0.5. As shown in Fig. 4.29 their φ'_p-values fit very well the values predicted by equ. (4.4).

The arrows in Figures 4.28 and 4.29 show the interrelation between these to diagrams, since the stress and density dependence of the material behaviour can hardly be demonstrated in one single diagram.

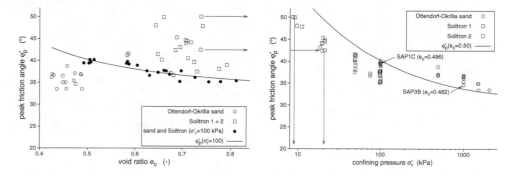

Fig. 4.28: Peak friction angles φ'_p of triaxial tests on Ottendorf-Okrilla sand and Soiltron in dependence of e_c; tests with σ'_r=100 kPa are marked with •, line after equ. (4.4)

Fig. 4.29: Peak friction angles φ'_p of triaxial tests on Ottendorf-Okrilla sand and Soiltron in dependence of σ'_r; line after equ. (4.4) exemplarily shown for $e_c = 0.5$

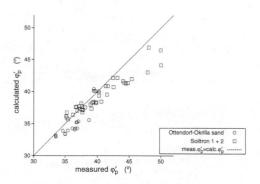

Fig. 4.30: Peak friction angles φ'_p of the triaxial tests and after equ. (4.4)

4.2.2   Normalisation of $\varepsilon_{a,p}$

Following relation between axial peak strain $\varepsilon_{a,p}$, stress and void ratio e_c was chosen to model the experimental results:

$$\varepsilon_{a,p}(\sigma_r', e_c) = \alpha_\varepsilon \cdot \left(\frac{\log\left(10\sigma_{ref}/\sigma_r'\right)}{e_c} \right)^{\beta_\varepsilon} . \qquad (4.5)$$

The factors were estimated with regression analysis in GNUPLOT as $\alpha_\varepsilon = 1.050$ and $\beta_\varepsilon = -1.820$, based on the measured values of all tests with 100 kPa cell pressure over a wide range of void ratios. The calculated values deviate from the measurements by less than $\Delta\varepsilon_{a,p} = \pm 1\%$ in almost all tests, see Fig. 4.33.

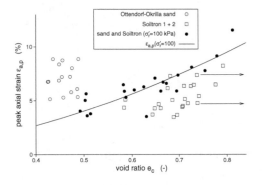

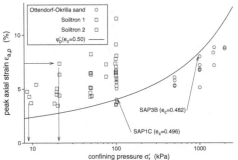

Fig. 4.31: Axial peak strain $\varepsilon_{a,p}$ of triaxial tests on Ottendorf-Okrilla sand and Soiltron in dependence of e_c; tests with σ_r'=100 kPa are marked with •, line after equ. (4.5)

Fig. 4.32: Axial peak strain $\varepsilon_{a,p}$ of triaxial tests on Ottendorf-Okrilla sand and Soiltron in dependence of σ_r'; line after equ. (4.5) exemplarily shown for $e_c = 0.5$

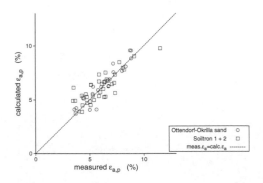

Fig. 4.33: Axial peak strain $\varepsilon_{a,p}$ of the triaxial tests and after equ. (4.5)

4.2.3   Normalisation of ψ'_p

The influence of barotropy and pyknotropy on the peak dilatancy angle of Ottendorf-Okrilla sand and Soiltron can be modelled with following equation:

$$\psi'_p(\sigma'_r, e_c) = \alpha_\psi \cdot \left(\frac{\log\left(10\sigma_{ref}/\sigma'_r\right)}{e_c} \right)^{\beta_\psi} , \qquad (4.6)$$

with $\alpha_\psi = 2.1 \cdot 10^{-4}$ and $\beta_\psi = 3.905$ obtained from a regression analysis based on the results of triaxial tests with 100 kPa confining pressure over a wide range of void ratios, as marked in Fig. 4.34. The deviation of the calculated values is $\Delta\psi'_p = \pm 2°$, see Fig. 4.36.

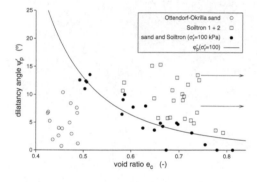 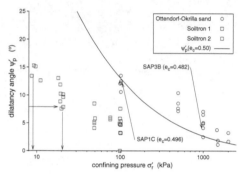

Fig. 4.34: Dilatancy angle ψ'_p of triaxial tests on Ottendorf-Okrilla sand and Soiltron in dependence of e_c; tests with σ'_r=100 kPa are marked with •, line after equ. (4.6)

Fig. 4.35: Axial peak strain $\varepsilon_{a,p}$ of triaxial tests on Ottendorf-Okrilla sand and Soiltron in dependence of σ'_r; line after equ. (4.6) exemplarily shown for $e_c = 0.5$

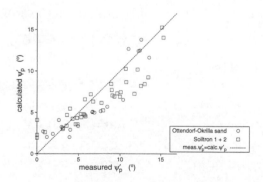

Fig. 4.36: Axial peak strain $\varepsilon_{a,p}$ of the triaxial tests and after equ. (4.6)

4.2.4   Requirements for Soiltron – an example

Solving the functions (4.4)-(4.6) for e_c we can calculate the density of the model soil which exhibits the same characteristics $(\varphi_p, \psi, \varepsilon_{a,p})$ as the soil in the prototype. We get different required densities to simulate the prototype behaviour in the $1g$ small scale model after equations (4.4)-(4.6). As was described in section 3.3.4, where the difficulties to *measure* the in-situ density were demonstrated, also in-situ a range of densities can be assumed. Therefore, variations in the density have to be accepted in a certain range. Consider, e.g., the relative density determination from SPT-correlations, as by Tavenas and LaRochelle[3]. There, the relative error of the N-value is $\pm 25\%$. Note that, the overall error on the resulting density depends on both, the N-value and the applied correlation. Tavenas and LaRochelle identified an overall error for at least $\pm 13\%$ for in-situ obtained relative densities.

Therefore, a deviation of the *model* density from the values calculated with equations. (4.4)-(4.6) of the aforementioned $\pm 13\%$ seems acceptable in the author's opinion.

To demonstrate the application of Soiltron to a small scale model test the geotechnical problem of shallow tunnelling (vertical cover H to diameter D ratio of $H/D = 1.65$) was chosen. The tunnel may be constructed in Ottendorf-Okrilla sand ($\gamma = 19$ kN/m^3, $e = 0.45$), Fig. 4.37, by using a tunnel boring machine with earth-pressure balance shield. The pressure at the heading face has to be adjusted in such a way that heaving and settlements are avoided as far as possible. The model test is described in detail in section 5.3. The density of the soil model is derived with equations (4.4)-(4.6).

As the dimensions of the model box and the model tunnel (see Fig. 4.38), and the depth of the model tunnel are fixed, only tunnels with $H/D = 1.65$ can be realised, see table 4.7. In this table 6 prototypes are summarised, and models with various geometrical reduction factors. For the models the ranges of void ratio for the soil model according to equations (4.4)-(4.6) are given. Furthermore, an average void ratio for the soil model and the deviation from the required void ratio, which should generate similar soil mechanical behaviour, are given.

The diagrams in Fig. 4.39 show extrapolations of equations (4.4)-(4.6) for

---

[3]Tavenas, F., LaRochelle, P.: Accuracy of relative density measurements, *Géotechnique*, 4 (**22**) 1972, pp. 549-562

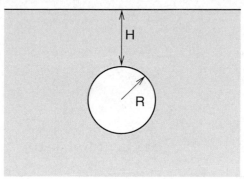

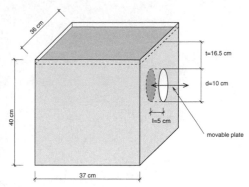

Fig. 4.37: Shallow tunnel

Fig. 4.38: Experimental box with model tunnel for the measurement of the pressure at the tunnel heading face; simplified sketch, details see section 5.3

	Prototype (Ottendorf-Okrilla sand)			Model (Soiltron)			
	Diameter D (m)	Depth H (m)	Stress level $(\sigma = \gamma z)$ (kPa)	Scale factor	Void ratio (range) Fig. 4.39	Void ratio (chosen)	Max. deviation
A	0.1	0.165	3	1:1	–	0.45	–
B	0.5	0.83	16	1:5	0.51-0.62	0.57	±9.7%
C	1.0	1.65	31	1:10	0.55-0.74	0.65	±14.7%
D	2.0	3.30	63	1:20	0.59-0.89	0.74	±20.3%
E	5.0	8.25	157	1:50	0.65-1.18	0.92	±29.0%
F	10.0	16.50	314	1:100	0.71-1.5	1.11	±35.7%

Tab. 4.7: Stability of tunnel heading face: modeled prototypes ($\gamma = 19$ kN/m³, $e = 0.45$)

given prototype situations. The soil state in the prototype situation (characterised by a stress σ and a density e) and the required characteristics for the model are marked in the diagrams.

When preparing the soil model with the average density calculated from equations (4.4)-(4.6) only examples (A), B and C appear to be feasible with this experimental equipment.

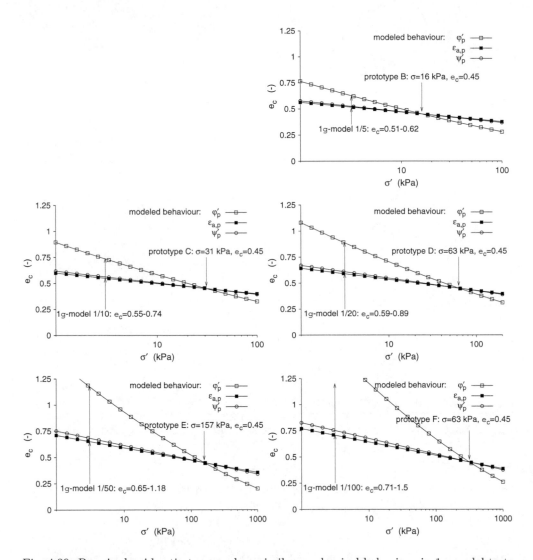

Fig. 4.39: Required void ratio to reproduce similar mechanical behaviour in 1g-model tests of a tunnel $H/D = 1.65$ in the model box, see section 5.3

4.3   Limitations of Soiltron

The main disadvantages of Soiltron are that it cannot be used for water-saturated soil models and with the pluviation under water. Due to their low density the additives float on water and when using pluviation the additive particles fall slower than the heavier soil particles which results in segragation of sand and additives. The flooding of a soil model is in the author's opinion possible.

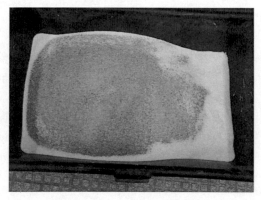

Fig. 4.40: Segregation of the additives after uncontrolled discharge. The white particles are polystyrene beads.

Polystyrene can be assumed for the duration of one model test as watertight, as the water absorption after being 28 days submerged in water is less than 3 Vol.%, see section 2.2.3. The surface of the polystyrene beads is watertight.

Dry Soiltron segregates easily during soil model preparation operations. A method to overcome this segregation is described in section 5.1. The soil models made with this method are evidently homogeneous mixtures.

Fig. 4.41: Surface of the soil model for the investigation of stability of tunnel heading face (see section 5.3), right detail of the soil model surface

Chapter 5

Applications

Examples of the application of Soiltron in this study are footing tests and the stability of a tunnel heading face. The footing on soil is apparently the simplest example for model tests, and of primary importance for soil engineering. The model tests to evaluate the support pressure on the tunnel heading face was chosen because this model test was, according to the author's inquiries, never done before. The presented results can be possibly used for comparison with centrifuge tests. The interested centrifuge experimentator is invited to repeat these tests.

Nevertheless, the model tests of this chapter should be only seen as demonstration models, which show the use of Soiltron — which originates from a theoretical approach — in practical applications.

5.1   Homogenisation of Soiltron

The mixture of the sand and the additives was done by hand in a bucket using a shovel and with the help of 99% methyl alcohol[1] (CH_3OH, boiling point 64.6°C, flash point 11°C, autoignition 245°C) or 95% ethyl alcohol (C_2H_5OH, boiling point 78.4°C, flash point 17°C, autoignition 425°C) which is less harmful, respectively, in the concentration $w \approx 1\%$. The alcohol improves the miscibility of sand and the additives due to a marginal capillary cohesion of the particles. In addition it improves the preparation of very loose samples, as can be seen e.g. with the oedometric compression tests of section 2.3. The

---

[1] This hint was given by A.Ph. Revuzhenko and V.P. Kosykh of the Solid Body Deformations Laboratory in the Mining Institute of the Siberian Branch of the Russian Academy of Sciences, Novosibirsk. I thankfully acknowledge this hint.
Attention: Methyl alcohol may cause respiratory tract irritation, is toxic if inhaled, may cause skin irritation and may be harmful if absorbed through the skin!
Therefore, for larger model tests ethyl alcohol should be used.

alcohol evaporates very fast from the specimens. Thus, it can be assumed that capillarity was completely removed after sample preparation..

Machined mixtures of soil and perlite are not advisable because perlite is very brittle and would pulverise. With polystyrene these mixtures should be possible. This additive is, therefore, more appropriate for larger models.

5.2 Footing tests

The applicability of Soiltron for $1g$-model tests was evaluated in a simple soil engineering problem: footing tests. As prototype the tests by Leussink et al.[2] were chosen. The prototypes are 1×1 m footings on dense Rheinsand ($I_n = 0.83$-0.86) and very dense Rheinsand ($I_n = 1.01$-1.18), see section 1.2.2 and Appendix A.2.

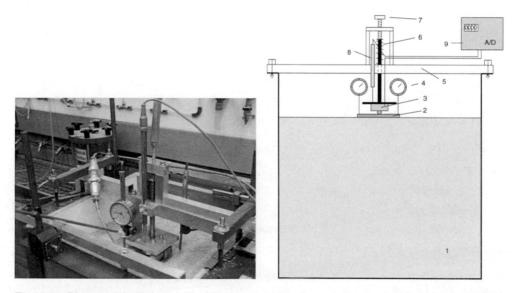

Fig. 5.1: Photograph and sketch of the experimental setup for the footing tests: 1 model box and model soil, 2 aluminium alloy plate 10×10 cm, 3 electric load cell, 4 dial gauges on two corners of the footing, 5 cross head, 6 driving rod with spring, 7 screw, 8 LVDT, 9 A/D-converter for load cell and LVDT measurements

---

[2]Leussink, H., Blinde, A., Abel, P.-G.: Versuche über die Sohldruckverteilung unter starren Gründungkörpern auf kohäsionslosem Sand (Tests to evaluate contact pressure of rigid foundations on cohesionless sand), Publications of the Institute of Soil and Rock Mechanics, University of Karlsruhe, Vol. 22, 1966

5.2.1   Experimental equipment

The model tests were done in an approximately cubical box with dimensions $l/w/h = 36/37/40$ cm. The sidewalls were plastic-coated chip-boards. A photography and a sketch of the experimental setup are shown in Fig. 5.1. The footing was simulated with a free moving quadratic aluminium alloy platen (10×10 cm). The soil model was 30 cm thick. The depth/width ratio is 3, which should be sufficient for single footings. The clearance to the sides is with approximately 1.5× the footing width too small to simulate bearing capacity tests. Note, that these were only preliminary investigations of the appropriateness of Soiltron. The model footing was aligned using a precision water level (maximal inclination ±0.1 mm/1 m).

The soil model was built in 6 layers, each with 5 cm height, with the undercompaction method and an undercompaction ratio of 1% using Ottendorf-Okrilla sand or Soiltron, respectively. The density for the model test with pure sand was chosen approximately equivalent to the densities given by Leussink *et al.*[2]. This test was done to show that the load-settlement characteristic predicted from the model test is unrealistic due to the high dilatancy of the pure sand under low pressure level.

Because the Rheinsand used by Leussink *et al.* is different from Ottendorf-Okrilla sand, the model footing test results can only be similar.

Rheinsand 0/5 mm consists of rounded grains. The minimum and maximum void ratios are $e_{min} = 0.44$ and $e_{max} = 0.78$. The friction angle is given as $\varphi = 34\text{-}38°$ (depending on test method and density). The cohesion is $c = 0\text{-}25$ kN/m^2. In comparison to Ottendorf-Okrilla sand, $e_{min} = 0.42$ and $e_{max} = 0.75$ are comparable, the peak friction angle was measured between 33.5° to 40.2°. The grain size of Ottendorf-Okrilla sand is 0/2 mm.

In Chapter 6 it is briefly described how to determine the density of Soiltron for specific tasks, such that the pressure-dependent relative density is similar to that of the prototype situation. Thus, Soiltron has similar soil mechanical behaviour at an n-times reduced pressure. But, what is the pressure in the soil in this problem? The average contact pressure under the footing at failure, or the average pressure in the soil at a certain depth under the footing during the loading or before loading of the footing?

In this specific case it is difficult to assess the required density of Soiltron, because (i) the pressure is applied to the system due to external forces, and (ii) Rheinsand was not available for the mixture with additives. Therefore, it was decided to conduct several model tests and to compare them with one 'good' prototype test, to estimate the required density for a similar load-settlement behaviour which can subsequently be used for additional tests, say with various embedment depths.

5.2.2 Model and prototype tests

For comparison, Leussink *et al.*'s test A_1/IX was chosen. The 1×1 m footing was placed on very dense Rheinsand ($I_n = 1.18$) with an embedment of $t = 0.0$ m. The original results are shown in Fig. A.4 of Appendix A.2. Fig. 5.2 shows the digitized results of this test and the average of the measurements.

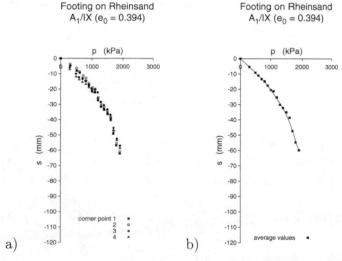

Fig. 5.2: Load-settlement results of Leussink *et al.*'s test A_1/IX; a) measurements on the 4 corner points, b) average of the load-settlement measurements and regression line of the average values

The following scaling factors apply to the 1:n model: length $1/n$, mass density 1, stress $1/n$, force $1/n^3$. In Fig. 5.3 measurements are shown in a model test on dense Ottendorf-Okrilla sand. The settlements were measured at two corners and at the driving rod (see Fig. 5.1) which corresponds to the settlement at the center of the footing. The fourth diagram shows the average of the three measurements. The measurements are shown in prototype scale. The load-settlement curve exhibits a peak which can be attributed to the high dilatancy of the dense soil.

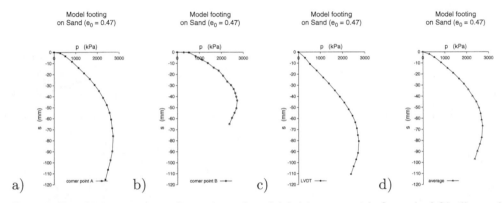

Fig. 5.3: Results in prototype dimensions of model footing test with Ottendorf-Okrilla sand at $e_0 = 0.47$: a) and b) measurements at two footing corners, c) measurements at the driving rod (circa at the center of footing), and d) average of the measurements a)-c)

Tests on soil models with lower density were conducted using Soiltron where polystyrene beads were added. Fig. 5.4 shows results of a test on Soiltron 1 (0.1 ml PS/1 g sand) with an initial void ratio $e_0 = 0.65$.

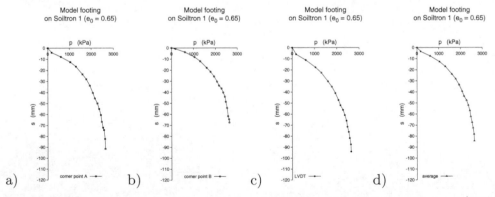

Fig. 5.4: Results in prototype dimensions of model footing test with Soiltron 1 at $e_0 = 0.65$: a) and b) measurements at two footing corners, c) measurements at the driving rod (circa at the center of footing), and d) average of the measurements a)-c)

Fig. 5.5 shows results of a test on Soiltron 1H (0.15 ml PS/1 g sand) with an initial void ratio $e_0 = 0.75$.

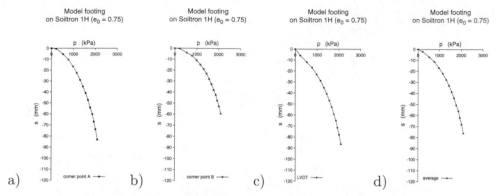

Fig. 5.5: Results in prototype dimensions of model footing test with Soiltron 1H at $e_0 = 0.75$: a) and b) measurements at two footing corners, c) measurements at the driving rod (circa at the center of footing), and d) average of the measurements a)-c)

5.2.3 Comparison of test results

All test results are summarised in Fig. 5.6. Comparing these tests, one can see that if the model soil density is decreasing ($b \rightarrow d$) the settlements at the same pressure are increasing and the ultimate load is decreasing. This can be attributed to pyknotropy (see Fig. 1.3, p. 3): decreasing stiffness and strength with decreasing density. Fig. 5.6d) shows good match with the prototype measurements. More tests with variation of embedment depth could be done with this density. These are intended for a project work of one of the author's students.

From the tests it can be seen that Soiltron is an adequate material for small-scale model tests. The model tests can be done very easy in short time, and it is possible to change boundary conditions.

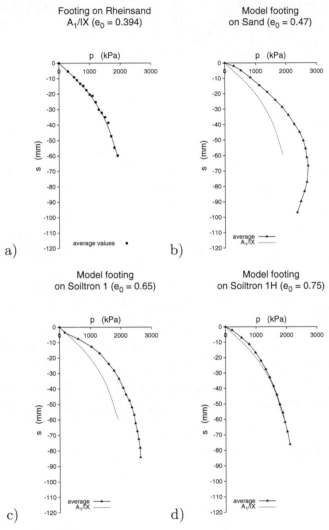

Fig. 5.6: a) Prototype; Model tests: b) sand at $e_0 = 0.47$, c) Soiltron 1 at $e_0 = 0.65$, d) Soiltron 1H at $e_0 = 0.75$

5.3   Stability of a tunnel heading face

5.3.1   Experimental equipment

Boundary conditions

The sidewalls of the experimental box were made of chip-boards with a smooth synthetic coating. The frictional coefficient of the sand on that material

$$\mu \approx 0.4 \quad (\approx \tan 22°)$$

was measured in direct shear tests using a 10×10 cm sand specimen ($\gamma_d = 16.0$ kN/m^3, $e_0 = 0.65$) with 21 mm thickness, see Fig. 5.7 and 5.8.

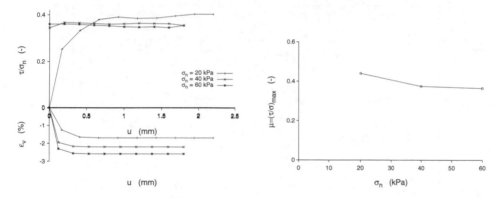

Fig. 5.7: Direct shear test for the evaluation of boundary friction of the model box

Fig. 5.8: Stress dependence of the frictional coefficient μ

With this boundary friction the silo effect had to be taken into account. JANSSEN's silo equation for unstressed surfaces[3]

$$\sigma(z) = \frac{\gamma r}{2K_0\mu} \left(1 - e^{-2K_0\mu z/r}\right) \quad , \tag{5.1}$$

leads to the vertical stress distribution as can be seen in Fig. 5.9, where the silo effect is approximately 12% at the crown of the model tunnel and 18% at the invert, see table 5.1 (for $\varphi = 40°$, according to triaxial test results of section 4.1 for pressures prevailing approximately in the model test). The maximum vertical stress due to own-weight in an infinitely high box of this size would be max $\sigma_v \approx 9.7$ kPa.

---

[3]Kolymbas, D.: Geotechnik – Tunnelbau und Tunnelmechanik, Springer Verlag Berlin, 1998

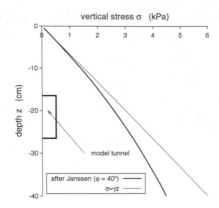

Fig. 5.9: Vertical stress distribution in the model test box ($\mu = 0.4$, $K_0 = 1 - \sin\varphi$)

depth	vertical stress (after JANSSEN)	vertical stress (without silo effect)	error
z (cm)	$\sigma_{v,J}$ (kPa)	σ_v (kPa)	(%)
16.5 (crown)	2.18	2.48	11.7
21.5 (center)	2.74	3.23	14.9
26.5 (invert)	3.26	3.98	17.9

Tab. 5.1: Vertical stress at the tunnel, ($\gamma=15$ kN/m³, $\mu=0.4$, $\varphi = 40°$, $K_0 = 1 - \sin\varphi$)

Model box

The tunnel model is shown in Fig. 5.10. The heading face is simulated with a rigid plate ⌀10 cm (2), fixed on a rod (4) which is movable in the tunnel tube (3). The plate (2) is pushed against the model soil (1) through a driving piston (6). To minimize restraints the piston (6) is driven with a screw (9) against a spring. The rod is guided in a cross head (7) and protected against tilt (10). The screw is fixed separately (8). The load cell (5) and LVDT (11) measurements were monitored with an A/D-converter. The surface was monitored on three positions using displacement gauges (12).

The surface displacement gauges measured with an accuracy of 0.01 mm. The precision of the LVDT which measured the movement of the heading plate was approximately ±0.005 mm. The load cell was calibrated using the calibration signal which was supplied by the producer of the device. The

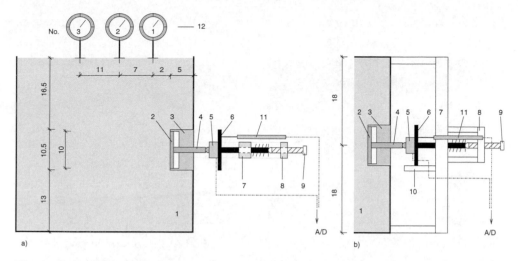

Fig. 5.10: Tunnel model: a) vertical section, b) detail in top view (dimensions in cm); 1 soil model and model box, 2 movable plate to simulate heading face, 3 tunnel, 4 rod, fixed on the heading face, 5 load cell, 6 driving piston with plate for fixing of the load cell and spring, 7 cross head, 8 fixing, 9 screw, 10 anti-twist device, 11 LVDT, 12 surface strain gauges

maximum payload of the load cell was 200 kg. In the tests, a maximum load of 148.8 kg was measured.

5.3.2 Test programme

5.3.2.1 General aspects

The model soil was built in by the hand tamping method, cf. section 1.4.4. The reasons for the use of this method are (i) the tunnel partly enters the box, therefore it would be impossible to pluviate sand under the tunnel, (ii) when using Soiltron, the light additive particles would fall slower than the sand particles, segregation would occur, (iii) the fluidisation method also had to be omitted, because the light additive particles would float on the water.

The preparation was done carefully using the undercompaction method with 1% undercompaction ratio in eight layers, see section 1.5.2. A predefined amount of sand or Soiltron was spread in the model box (Fig. 5.11a) and compacted by tamping on the surface. The levels of the intermediate surfaces were controlled over the whole area of the box (Fig. 5.11b). The plate to simulate the heading face was lubricated before every test with grease (Fig. 5.12a) and additionally covered with a latex membrane (Fig. 5.12c) after inserting it into the tunnel tube (Fig. 5.12b). The completed model test box is shown in Fig. 5.13a. The strain gauges for monitoring the soil surface are shown in Fig. 5.13b.

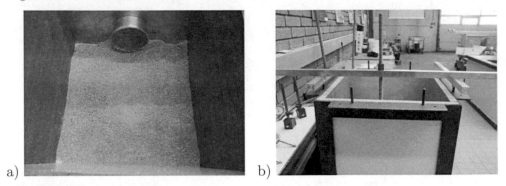

a)   b)

Fig. 5.11: Tunnel test: a) spreaded Soiltron in the model box before compaction, b) compactor: plate fixed on a height adjustable rod

a)   b)   c)

Fig. 5.12: Tunnel test: a) lubrication of the plate, b) tunnel, c) in tunnel inserted plate, which is additionally covered with a membrane to prevent grains from entering the gap between plate and tunnel tube

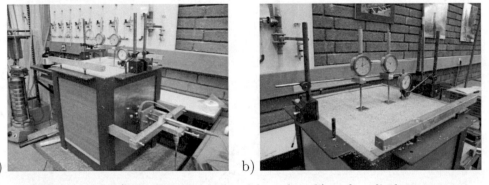

a)   b)

Fig. 5.13: Tunnel test: a) completed tunnel model test box, b) surface displacement gauges

Since these tests were performed in a box with a fixed tunnel, the heading face is already stressed after filling of the model box. This pre-stress is varying according to the applied compaction energy, although undercompaction method was used.

With the tunnel equipped as described above, model tests to evaluate the support pressure at the heading face to avoid heaving and settlement are possible.

5.3.2.2   Support pressure to avoid heave

Model A: after Tab. 4.7

A tunnel $D = 0.1$ m in a depth $H = 0.165$ m is reproduced in a 1:1-laboratory model. The model soil was made of Ottendorf-Okrilla sand with a void ratio $e = 0.47$ ($\gamma = 17.93$ kN/m^3, $r_e \approx 0.15$), higher densites were not possible with dry sand. Model A was the first one of this test series. Here, the surface heaving was measured only at one position of the soil model at 9 cm distance from the tunnel face. In the other model tests the heaving was measured at three positions (Fig. 5.10). Fig. 5.14 shows the average pressure Q (quotient of measured force F at the plate and area of the plate A) at the heading face vs. the penetration s of the rigid plate into the soil and the heaving h in dependence of s. The pressure seems to tend to a limit, which was not reached in the test. In a further test this limit has been passed, see Fig. 5.16. The surface heaving is almost linearly increasing with s from the beginning of penetration of the plate into the soil. Fig. 5.15 shows the surface heaving h in dependence of the pressure Q at the heading face. The plate was pre-stressed due to the compaction of the soil model, in this test: $Q_0 = 1.4$ kPa, which corresponds to an average horizontal pressure (cf. table 5.1) of $\sigma_h = \gamma h K_0 = 17.93$ kN/m$^3 \cdot 0.215$ m $\cdot 0.36 = 1.38$ kN/m^2 (for $K_0 = 1 - \sin\varphi$ and $\varphi = 40°$).

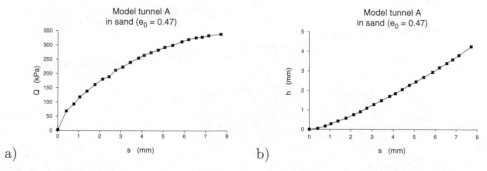

a)   b)

Fig. 5.14: Measurements on 1:1-model A: a) average pressure Q which is required to press in the plate into the soil at distance s, b) surface heaving h due to press in operation

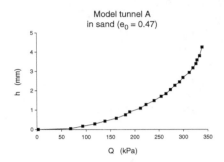

Fig. 5.15: Measurements on 1:1-model A: surface heaving in dependence of pressure

The response of the the soil due to displacement of the plate is shown in Fig. 5.16. The pressure decreases drastically while due to the press-in operations the heavings increases. After the test a distinct discontinuity was visible on the surface.

The plate which simulates the tunnel heading face has a thickness of 10 mm. The tests were stopped when the plate penetrated into the soil by approximately 8 mm.

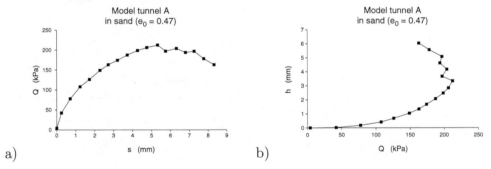

a)   b)

Fig. 5.16: Results of 1:1-model A with collapse: a) pressure Q in dependence of press in distance, b) surface heaving in dependence of pressure (h measured only at one position)

Model B: after Tab. 4.7

The prototype situation of a tunnel $D = 0.5$ m in depth $H = 0.83$ m was $1 : 5$ modelled in a model soil (pure Ottendorf-Okrilla sand) with void ratio $e = 0.57$. The heaving was measured at three positions at the surface of the soil model, as shown in Fig. 5.10.

Fig. 5.17 shows the average pressure Q at the heading face vs. the penetration s of the rigid plate into the soil, and the heaving h in dependence of s. Fig. 5.15 shows the surface heaving h in dependence of the pressure Q at the heading face. The initial stress at the plate was 0.8 kPa in model dimension – smaller than in model A. Collapse of the soil could not be identified in the plots, neither by visual inspection of the model surface after the test.

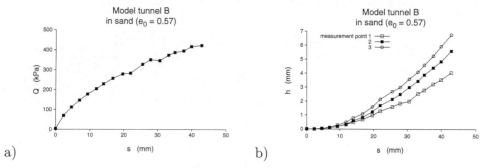

a)   b)

Fig. 5.17: Measurements on 1:5-model B (prototype dimensions): a) average pressure Q which is required to press in the plate into the soil at distance s, b) surface heaving h due to press in operation

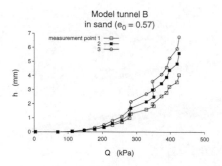

Fig. 5.18: Measurements on 1:5-model B: surface heaving in dependence of pressure

Model C: after Tab. 4.7

The prototype situation of a tunnel $D = 1.0$ m in depth $H = 1.65$ m was
$1 : 10$ modelled in a model soil (Soiltron 1) with void ratio $e = 0.65$. As before, the measurements were processed into similar diagrams, see Figures 5.19
and 5.20. The dimensions of Q, s, h are in prototype scale. Before the test, the
force at the plate was measured: $Q_0 = 1.6$ kPa (in model scale) – comparable
with model A.

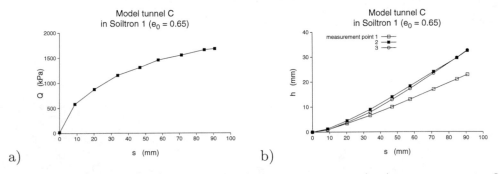

a)   b)

Fig. 5.19: Measurements on 1:10-model C (prototype dimensions): a) average pressure Q
which is required to press in the plate into the soil at distance s, b) surface heaving h due
to press in operation

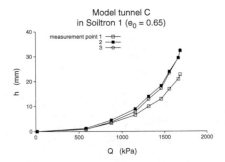

Fig. 5.20: Measurements on 1:10-model C: surface heaving in dependence of pressure

5.3.2.3 Support pressure to avoid settlements

These tests required adaptations of the existing test box, which are still in progress.

5.3.2.4 Discussion of the results

Soiltron was used in the tests to assess heavings of the ground surface due to shallow tunnelling with EPB-shields, where the heading face is stabilized with pressurized slurry. In model tests, three different situations were simulated. In all models, same prototype conditions were assumed: Ottendorf-Okrilla sand $\gamma = 19 \text{ kN/m}^3$, $e = 0.45$. Model A was simulated in the geometrical scale 1:1, model B 1:5 and model C 1:10. The density of the model ground was chosen according to the normalization routines of section 4.2. For models A and B it was possible to use the prototype sand in the model, whereas model C had to be prepared with a very low density, which was collapsible when made of pure sand. Therefore, Soiltron was used for the preparation of the model.

In all models, it can be seen that even relatively low pressures on the heading face generate heavings on the surface. Prototype measurements should be done.

Clearly, in reality the pressure on the ground is due to air or slurry pressure, not due to a rigid platen. In further studies the heavings could be investigated with, say, a rubber membrane which presses against a model soil made with Soiltron.

Chapter 6

Suggestions for the application of Soiltron and for further investigations

Soiltron has proved to be an adequate soil for $1g$ small scale model tests. The mechanical behaviour of the original material – Ottendorf-Okrilla sand – was appropriately influenced by the addition of the soft and light particles polystyrene and perlite.

Based on long period of experimental investigations, the author's suggestions for the use of Soiltron are:

1. **Find out the mechanical behaviour of the prototype soil.**

    Triaxial tests proved to be a good tool for the testing in various stress paths. The soil behaviour is affected by pressure and density. Therefore, the tests should cover a wide range of pressures, at least the region which is interesting in the prototype situation.

    If no triaxial test results are available, say 25 tests – each five with comparable initial void ratio at five pressures – should be sufficient if the tests are made with sophisticated instruments in a reliable apparatus, not as shown on Fig. 6.1 from Menzies[1].

2. **Relate the obtained behaviour to your model.**

    Simple regression analysis showed in the case of Ottendorf-Okrilla sand a way to calculate the required density of the model soil.

---

[1]Menzies, B.K.: Applying Modern Measures, Article featured in Ground Engineering magazine, July 1997

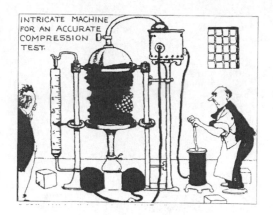

Fig. 6.1: "Applying Modern Measures", Menzies[1]

3. **Define the model.**

   Soiltron is not the universal remedy! In the case of here used materials the found relations of soil characterising parameters and influencing variables had the disadvantage that they were not unique. This means, that a certain Soiltron has only a limited range of applicability and should, therefore, be thoroughly taylored.

   Likewise, soil density in-situ has a large scatter and the possibilities for site-investigations and conclusions from these measurements are limited.

4. **Apply!**

   Interesting phenomena can be investigated in model tests. Not only qualitative measurements should be taken into account. Often, these models can help to verificate numerical results.

Presently, model tests are conducted in the Institute for Geotechnical and Tunnel Engineering (University of Innsbruck) for the investigation of settlements due to shallow tunnelling, which is caused although the tail gap between tunnel lining and ground is filled with grout. In these tests Soiltron was used as model soil by M. Mähr. The publication of the test results is envisaged for 2005.

So far, only one soil was investigated for the application as Soiltron. In further studies other soils should be also investigated.

Soiltron should be applied also to other model tests, since it is a capable material for many model investigations.

Chapter 7

Summary

In this thesis a new approach for $1g$ model tests was investigated. Herein, the similarity in the mechanical behaviour of soils due to simulation of barotropy by pyknotropy was used. Model tests are a good alternative to large scale tests due to their lower cost and the possibility to control the test conditions continuously.

For that reason the mechanical behaviour of a 'prototype' material was investigated in detail in triaxial tests using sophisticated devices, such as hydraulic pressure control devices, local strain gauges, remote measurement of deformation based on the PIV method.

The mechanical similarity of model and prototype was achieved through addition of soft and light particles to the prototype soil. The barotropy (stress dependence) of the soil behaviour was simulated with pyknotropy (density dependence). It was shown that the soil behaviour is similar at different pressures when the stress dependent relative densities are similar. Simple relations were found for the estimation of required densities in dependence of pressure to generate desired peak friction angles, peak strains and dilatancy angles at the peak states of the stress-strain behaviour of the investigated soil. These relationships offer a handy tool for experimentators.

The application of Soiltron was shown in two demonstration model tests. Model test for the investigation of the load-settlement characteristic of a quadratic footing on sand and the upheaval of the ground surface due to shallow tunnelling using an earth-pressure balance shield were conducted. The model footing test results were compared with large scale tests under laboratory conditions, whereas the tunnel tests were done without comparison due to lack of reports about field measurements.

References

Adalier, K., Elgamal, A., Meneses, J., Baez, J.I.: Stone columns as liquefaction countermeasure in non-plastic silty soils, *Soil Dynamics and Earthquake Engineering*, 7 (**23**) 2003, pp. 571-584

Adrian, R.J.: Dynamic ranges of velocity and spatial resolution of particle image velocimetry, *Measurement Science and Technology*, 12 (**8**) 1997, pp. 1393-1398

Al-Akel, S.: Modellversuche an einer im Boden eingespannten Wand (Model tests on a wall fixed in soil), Ohde-Colloquium 2001, in Pubblications of the Institute for Geotechnics, Technical University of Dresden, Vol. 9, 2001

Ali, S.R., Pyrah, I.C., Anderson, W.F.: A novel technique for the evaluation of membrane penetration, *Géotechnique* 3 (**45**) 1995, pp. 545-548

Alshibli, K.A., Sture, S., Costes, N.C., Frank, M., Lankton, M., Batiste, S., Swanson, R.: Assessment of Localized Deformations in Sand Using x-ray Computed Tomography, ASTM, *Geotechnical Testing Journal*, 3 (**23**) 2000, pp. 274-299

Andersen, K.H.: Panel discussion to "The use of physical models in design", in *Design Parameters in Geotechnical Engineering*, Proceedings of the VIIth European Conference on Soil Mechanics and Foundation Engineering, Brighton/England 1979, British Geotechnical Society, London, 1980, Vol. 4, pp. 315-317

Arnold, A.: Modellversuche zum Erddruck auf Winkelstützwände (Model tests to the earth-pressure on buttress), Ohde-Colloquium 2001, in Pubblications of the Institute for Geotechnics, Technical University of Dresden, Vol. 9, 2001

Athanasopoulos, G.A., Pelekis, P.C., Xenaki, V.C.: Dynamic properties of EPS geofoam: An experimental investigation, *Geosynthetics International*, 3 (**6**) 1999, pp. 171-194

Atkinson, J.: An Introduction to the Mechanics of Soils and Foundations, McGraw-Hill Book Company, London 1993

Baecher, G.B., Christian, J.T.: Reliability and Statistics in Geotechnical Engineering, John Wiley & Sons Ltd., Chichester, 2003

Baldi, G., Nova, R.: Membrane penetration effects in triaxial testing, *Journal of Geotechnical Engineering*, 3 (**110**) 1984, pp. 403-420

Baldi, G., Hight, D.W., Thomas, G.E.: A Reevaluation of Conventional Triaxial Test Methods, in *Advanced Triaxial Testing of Soil and Rock*, ASTM STP 977, Donaghe, Chaney and Silver, Eds., ASTM Philadelphia 1988, pp. 219-263

Bishop, A.W., Henkel, D.J.: The Measurement of Soil Properties in the Triaxial Test, Second Edition, Edward Arnold (Publishers) Ltd., London, 1962

Boháč, J., Feda, J.: Membrane penetration in triaxial tests, *Geotechnical Testing Journal*, 3 (**15**) 1992, pp. 288-294

Bransby, M.F., Springman, S.M.: 3-D Finite Element Modelling of Pile Groups Adjacent to Surcharge Loads, *Computers and Geotechnics*, 4 (**19**) 1996, pp. 301-324

Bransby, M.F., Yun, G.J.: Centrifuge investigation of the horizontal capacity of shallow footings on sand, Proceedings ISOPE 2003, Honolulu, USA, May 2003

Bucky, P.B.: Use of models for the study of mining problems, Technical Publication No. 425,

American Institute of Mining and Metallurgical Engineers, 1931, pp. 3-28

Bucky, P.B., Fentress, A.L.: Application of principles of similitude to design of mine workings, Technical Publication No. 529, American Institute of Mining and Metallurgical Engineers, 1934, pp. 3-20

Bucky, P.B., Solakian, A.G., Baldin, L.S.: Centrifugal method of testing models, *Civil Engineering*, 5 (**5**) 1935, pp. 287-290

Cenedese, A., Pocecco, A. Querzoli, G: Effects of image compression on PIV and PTV analysis, *Optics & Laser Technology*, 2 (**31**) 1999, pp. 141-149

Clough, H.F., Vinson, T.S.: Centrifuge model experiments to determine ice forces on vertical cylindrical structures, *Cold Regions Science and Technology*, 3 (**12**) 1986, pp. 245-259

Cresswell, A., Barton, M.E., Brown, R.: Determining the maximum density of sands by pluviation, *Geotechnical Testing Journal*, 4 (**22**) 1999, pp. 324-328

Desrues, J., Chambon, R., Mokni, M., Mazerolle, F.: Void ratio evolution inside shear bands in triaxial sand specimens studied by computed tomography, *Géotechnique*, 3 (**46**) 1996, pp. 529-546

Desrues, J., Viggiani, G.: Strain localization in sand: an overview of the experimental results obtained in Grenoble using stereophotogrammetry, *International Journal for Numerical and Analytical Methods in Geomechanics*, 4 (**28**) 2004, pp. 279-321

Dewoolkar, M.M., Ko, H.-Y., Stadler, A.T., Astaneh, S.M.F., A substitute pore fluid for seismic centrifuge modeling, Geotechnical Testing Journal, 3 (**22**), 1999, pp. 196-210

Di Florio, D., Di Felice, F., Romano, G.P.: Windowing, re-shaping and re-orientation interrogation windows in particle image velocimetry for the investigation of shear flows, *Measurement Science and Technology*, 7 (**13**) 2002, pp. 953-962

Doanh, T., Ibraim, E., Matiotti, R.: Undrained instability of very loose Hostun sand in triaxial compression and extension. Part 1: experimental observations, *Mechanics of Cohesive Frictional Materials*, 1 (**2**) 1997, pp. 47-70

Fellin, W., Kolymbas, D.: Hypoplastizität für leicht Fortgeschrittene (Hypoplasticity for the lower intermediate), *Bautechnik*, 12 (**79**) 2000, pp. 830-841

Finn, W.D.L., Fujita, N.: Piles in liquefiable soils: seismic analysis and design issues, *Soil Dynamics and Earthquake Engineering*, 9-12 (**22**) 2002, pp. 731-742

Fromm, H.: Experimentelle Überprüfung von Oberflächensetzungen infolge Ringspaltes und seiner Verpressung, MSc. thesis, Institute of Geotechnical and Tunnel Engineering, University of Innsbruck, 2002

Frost, J.D., Park, J.-Y.: A critical assessment of the moist tamping technique *Geotechnical Testing Journal*, 1 (**26**) 2003, pp. 57-70

Germaine, J.T., Ladd, C.C.: Triaxial Testing of Saturated Cohesive Soils, in *Advanced Triaxial Testing of Soil and Rock*, ASTM STP 977, Donaghe, Chaney and Silver, Eds., ASTM Philadelphia 1988, pp. 421-459

Grant, I., Smith, G.H.: Modern developments in Particle Image Velocimetry, *Optics & Lasers in Engineering*, 3-4 (**9**) 1998, pp. 245-264

Gudehus, G.: A comprehensive constitutive equation for granular material, *Soils & Foundations*, 1 (**36**) 1996, pp. 1-12

Heikkilä, J., Silvén, O.: A four-step camera calibration procedure with implicit image correction, Proceedings of the IEEE Computer Society Conference on Computer Vision and

Pattern Recognition (CVPR'97), San Juan, Puerto Rico, pp. 1106-1112

Herle, I.: Hypoplastizität und Granulometrie einfacher Korngerüste (Hypoplasticity and granulometry of simple granular structures), PhD Dissertation, University of Karlsruhe, 1997

Herle, I.: A relation between parameters of a hypoplastic constitutive model and grain properties, in Proc. of the Fourth International Workshop on Localization and Bifurcation Theory for Soils and Rocks, Gifu, Japan, 28. Sept.-2. Oct. 1997, A.A. Balkema, Rotterdam, 1998

Hettler, A.: Verschiebungen starrer und elastischer Gründungskörper in Sand bei monotoner und zyklischer Belastung (Displacements of rigid and elastic foundations in sand under monotonic and cyclic loading), Publications of the Institute of Soil and Rock Mechanics, University of Karlsruhe, Vol. 90, 1981

Hettler, A.:Horizontal belastete Pfähle mit nichtlinearer Bettung in körnigen Böden (Horizontal loaded piles with non-linear bedding in non-cohesive soils), Publications of the Institute of Soil and Rock Mechanics, University of Karlsruhe, Vol. 102, 1986

Hiegelsperger, M.: Nichtstandardmäßige Triaxialversuche (Non-standard triaxial tests), MSc. thesis at the University of Innsbruck, 2004

Hofstetter, G.: Skriptum zur Vorlesung Festigkeitslehre 1 (Script to the lecture 'Strength of materials 1'), University of Innsbruck, Vol. 1998/99

Holzhäuser, J.: Experimentelle und numerische Untersuchungen zum Tragverhalten von Pfahlgründungen im Fels (Experimental and numerical investigations to the bearing capacity of pile foundations in rock), Publications of the Institute and Experimental Station for Geotechnics, Technical University Darmstadt, Vol. 42, 1998

Huang, H., Dabiri, D., Gharib, M.: On errors of digital particle image velocimetry, *Measurement Science and Technology*, 12 (**8**) 1997, pp. 1427-1440

Jovanović, M.: Historische Holzgründungen – Tragverhalten in weichem Untergrund (Historical timber foundations – Bearing capacity in soft ground), PhD thesis, Publications of the Institute of Soil and Rock Mechanics, University of Karlsruhe, Vol. 153, 2002

Kamata, H., Mashimo, H.: Centrifuge model test of tunnel face reinforcement by bolting, *Tunnelling and Underground Space Technology*, 2-3 (**18**) 2003, pp. 205-212

Kiekbusch, M., Schuppener, B.: Membrane penetration and its effect on pore pressures, *Journal of the Geotechnical Engineering Division*, 11 (**103**) 1977, pp. 1267-1279

Kolymbas, D.: Geotechnik – Bodenmechanik und Grundbau, Springer-Verlag Berlin, 1998

Kolymbas, D.: Geotechnik – Tunnelbau und Tunnelmechanik, Springer Verlag Berlin, 1998

Kolymbas, D.: Introduction to Hypoplasticity, in *Advances in Geotechnical Engineering and Tunnelling*, Vol. 1, A.A. Balkema, Rotterdam, 2000

Kramer, S.L., Sivaneswaran, N.: Stress-path dependent correction for membrane penetration, *Journal of Geotechnical Engineering*, 12 (**115**) 1989, pp. 1787-1804

Lade, P.V., Hernandez, S.B.: Membrane penetration effects in undrained tests, *Journal of the Geotechnical Engineering Division*, 2 (**103**) 1977, pp. 109-125

Leussink, H., Blinde, A., Abel, P.-G.: Versuche über die Sohldruckverteilung unter starren Gründungkörpern auf kohäsionslosem Sand (Tests to evaluate contact pressure of rigid foundations on cohesionless sand), Publications of the Institute of Soil and Rock Mechanics, University of Karlsruhe, Vol. 22, 1966

Linemann, R., Runge, J., Sommerfeld, M., Weißgüttel: Densification of bulk materials in process engineering, in *Advances in Geotechnical Engineering and Tunnelling* Vol. 3, Kolymbas, D. & Fellin, W. (ed.), A.A.Balkema, Rotterdam, 2000

Linton, P.F., McVay, M.C., Bloomquist, D.: Measurements of Deformations in the Standard Triaxial Environment with a Comparison of Local versus Global Measurements on a Fine, Fully Drained Sand, in *Advanced Triaxial Testing of Soil and Rock*, ASTM STP 977, Donaghe, Chaney and Silver, Eds., ASTM Philadelphia 1988, pp. 202-215

Mayer, G.: Untersuchungen zum Tagverhalten von Verpreßankern in Sand (Investigations to the bearing capacity of injection anchors in sand), Reports of the Geotechnical Engineering Institute, Technical University Berlin, Vol. 12, 1983

Mélix, P.: Modellversuche und Berechnungen zur Standsicherheit oberflächennaher Tunnel (Model tests and calculations to the stability of shallow tunnels), Publications of the Institute of Soil and Rock Mechanics, University of Karlsruhe, Vol. 103, 1987

Menzies, B.K.: Applying Modern Measures, Article featured in Ground Engineering magazine, July 1997

Menzies, B.K.: Near-surface site characterisation by ground stiffness profiling using surface wave geophysics; Instrumentation in Geotechnical Engineering, H.C. Verma Commemorative Volume, Eds. K.R. Saxena and V.M. Sharma, Oxford & IBH Publishing Co. Pvt. Ltd., New Delhi, Calcultta, 2001, pp. 43-71

Mikasa, M., Takada, N.: Significance of centrifugal model tests in soil mechanics, Proceedings of the VIII ICSMFE, Vol. 1.2, pp. 273-278, Moscow, 1973

Mitchell, R.J.: The eleventh annual R.M. Hardy keynote address, 1997: Centrifugation in geoenvironmental practice and education, *Canadian Geotechnical Journal*, 5 (**35**) 1998, pp. 630-640

Molenkamp, F., Luger, H.J.: Modelling and minimization of membrane penetration effects in tests on granular soils, *Géotechnique*, 4 (**31**) 1981, pp. 471-486

Moo-Young, H., Myers, T., Tardy, B., Ledbetter, R., Vanadit-Ellis, W., Kim, T.-H.: Centrifuge simulation of the consolidation characteristics of capped marine sediment beds, *Engineering Geology*, 3-4 (**70**) 2003, pp. 249-258

Moreno, D., Mendoza Santoyo, F., Funes-Gallanzi, M., Fernandez Orozco, S.: An optimum data display method, *Optics & Laser Technology*, 2 (**32**) 2000, pp. 121-128

Newland , P.L., Allely, B.H.: Volume changes in drained triaxial tests on granular materials, *Géotechnique*, 1 (**7**) 1957, pp. 17-34

Nübel, K.: Experimental and numerical investigation of shear localization in granular material, PhD University of Karlsruhe, 2002, Vol. 159

Nübel, K., Weitbrecht, V.: Visualization of localization in grain skeletons with Particle Image Velocimetry, *Journal of Testing and Evaluation*, 4 (**30**) 2002, pp. 322-328

Okamoto, K., Nishio, S., Saga, T., Kobayashi, T.: Standard images for particle image velocimetry, *Measurement Science and Technology*, 6 (**11**) 2000, pp. 685-691

Olsen, M.G., Adrian, R.J.: Brownian motion and correlation in particle image velocimetry, *Optics & Laser Technology*, 2 (**32**) 2000, pp. 621-627

Ovesen, N.K.: Panel discussion to "The use of physical models in design", in *Design Parameters in Geotechnical Engineering*, Proceedings of the VIIth European Conference on Soil Mechanics and Foundation Engineering, Brighton/England 1979, British Geotechnical Society, London, 1980, Vol. 4, pp. 318-323

Plaßmann, B.: Zur Optimierung der Meßtechnik und der Auswertemethodik bei Pfahlin-tegritätsprüfungen (To the optimisation of measurement techniques and evaluation meth-ods for pile integrity tests), Publications of the Institute of Foundation Engineering and Soil Mechanics, Technical University Braunschweig, Vol. 67, 2002

Pokrovsky, G.I., Fedorov, I.S.: Studies of the soil pressures and soil deformations by means of an centrifuge, In A. Casagrande, P.C. Rutledge and J.D. Watson (eds), Proceedings of the First International Confonference ISSMFE (Harvard), Vol. I, p 70, 1936

Porbaha, A., Zhao, A., Kobayashi, M., Kishida, T.: Upper bound estimate of scaled rein-forced soil retaining walls, *Geotextiles and Geomembranes*, 6 (**18**) 2000, pp. 403-413

Raju, V.S., Sadasivan, S.K.: Membrane penetration in triaxial tests on sands, *Journal of the Geotechnical Engineering Division*, 4 (**100**) 1974, pp. 482-489

Ramana, K.V., Raju, V.S.: Membrane penetration in triaxial tests, *Journal of the Geotech-nical Engineering Division*, 2 (**108**) 1982, pp. 305-310

Randolph, M.F., House, A.R.: The complementary roles of physical and computational modelling, *International Journal of Physical Modelling in Geotechnics*, 1 (**1**) 2001, pp. 1-8

Rinawi, W.: Particle Image Velocimetry (PIV) applied on triaxial tests, MSc. thesis, Insti-tute of Geotechnical and Tunnel Engineering, University of Innsbruck, 2004

Roscoe,K.H., Schofield, A.N., Thurairajah, A.: An evaluation of test data for selecting a yield criterion for soils, *ASTM Special Technical Publication No. 361*, 1963, pp. 111-128

Rowe, P.W., Barden, L.: Importance of free ends in triaxial testing, *Journal of Soil Me-chanics and Foundation Division*, ASCE, 1 (**90**) 1964, pp. 1-27

Salomo, K.-P.: Pressenkräfte und Bodenverformungen beim Rohrvortieb (Driving forces and ground movements during pipe driving operations), Reports of the Geotechnical Engineering Institute, Technical University Berlin, Vol. 6, 1980

Sarsby, R.W., Kalteziotis, N., Haddad, E.H.: Compression of "free ends" in triaxial testing, *Journal of the Geotechnical Engineering Division*, 1 (**108**) 1982, pp. 83-107

Schanz, T., Vermeer, P.A.: Angles of friction and dilatancy of sand, *Géotechnique*, 1 (**46**) 1996, pp. 145-151

Schofield, A.N.: Cambridge Geotechnical Centrifuge Operations, *Géotechnique*, 2 (**30**) 1980, pp. 227-268

Sharma, J.S., Bolton, M.D.: A New Technique for Simulation of Collapse of a Tunnel in a Drum Centrifuge, Technical Report No. CUED/D-SoildTR286, Cambridge University Engineering Department, August 1995

Sharma, J.S., Bolton, M.D.: Centrifuge Modelling of an Embankment on Soft Clay Rein-forced with a Geogrid, *Geotextiles and Geomembranes*, 1 (**14**) 1996, pp. 1-17

Sladen, J.A., Handford, G.: A potential systematic error in laboratory testing of very loose sands, *Canadian Geotechnical Journal*, 3 (**24**) 1987, pp. 462-466

Sosna, K.: Pevostní charakteristiky směsí písku (Stiffness charakteristics of sand mixtures), MSc. thesis at the Charles University Prague, 2003

Stuit, H.G.: Sand in the geotechnical centrifuge, PhD thesis, Technical University Delft, 1995

Sveen, J.K.: An introduction to MatPIV v. 1.4, http://www..math.uio.no

Tatsuoka, F., Nakamura, S., Huang, C., Tani, K.: Strength Anisotropy and Shear Band

Direction in Plane Strain Tests of Sand, *Soils and Foundations*, 1 (**30**) 1990, pp. 35-54

Tavenas, F., LaRochelle, P.: Accuracy of relative density measurements, *Géotechnique*, 4 (**22**) 1972, pp. 549-562

Tejchmann, J.: Modelling of shear localisation and autogeneous dynamic effects in granular bodies, Publications of the Institute of Soil and Rock Mechanics, University Fridericiana of Karlsruhe, Vol. 140, 1997

Thiam-Soon, Tan: Two-phase soil study: A. Finite strain consolidation, B. Centrifuge scaling considerations, Technical Report, California Institute of Technology, 1985

Udrea, D.D., Bryanston-Cross, P.J., Lee, W.K., Funes-Gallanzi, M.: Two sub-pixel processing algorithms for high accuracy particle centre estimation in low seeding density particle image velocimetry, *Optics & Laser Technology*, 5 (**28**) 1996, pp. 389-386

Vaid, Y.P., Negussey, D.: Preparation of reconstituted samples, in *Advanced Triaxial Testing of Soil and Rock*, ASTM STP 977, Donaghe, Chaney and Silver, Eds., ASTM Philadelphia 1988, pp. 405-417

Weißenbach, A.: Reports on Measurement and Evaluations of measurements, in Series of the Department "Baugrund-Grundbau" University of Dortmund, Volumes 3-15 and 17-20, 1991-1994

White, D.J., Take, W.A., Bolton, M.D.: Measuring soil deformation in geotechnical models using digital images and PIV analysis, 10th International Conference on Computer Methods and Advances in Geomechanics, Tucson, Arizona 2001, Balkema, pp. 997-1002

v. Wolffersdorff, P.-A.: Verformungsprognosen für Stützkonstruktionen (Predictions of deformations of retaining structures), Publications of the Institute of Soil and Rock Mechanics, University Fridericiana in Karlsruhe, Vol. 141, 1997

Wu, W., Kolymbas, D.: Hypoplasticity then and now, in *Constitutive Modelling of Granular Materials*, Springer-Verlag Berin Heidelberg, 2000, pp. 57-105

Yang, Q.S., Poorooshasb, H.B.: Numerical Modeling of Seabed Ice Scour, *Computers and Geotechnics*, 1 (**21**) 1997, pp. 1-20

Yun, G, Bransby, M.F.: Centrifuge modeling of the horizontal capacity of skirted foundations on loose sand, Proceedings ICOF 2003, Dundee

Zeng, X.: Benefit of collaboration between centrifuge modeling and numerical modeling, in Proceedings of NSF International Workshop on Earthquake Simulation in Geotechnical Engineering, Cleveland, Ohio (USA), 8-10 November 2001

Modelling in Geotechnics, Course summer term 2003, ETH Zurich - Institute of Geotechnical Engineering, `http://geotec4.ethz.ch/mig/`, Chapter 7

Appendix A

Appendix

A.1 Relative densities of the specimens

The pressure dependent relative densities r_e after consolidation and prior the triaxial tests were:

SAP (Ottendorf-Okrilla sand)		Soiltron 1 (polystyrene)		Soiltron 2 (perlite)	
SAP1B	0.63	SO1P1A	0.45	SO2P1A	0.58
SAP1C	0.58	SO1P1B	0.62	SO2P1B	0.74
SAP1D	0.66	SO1P1E	0.58	SO2P1C	0.65
SAP1G	0.62	SO1P1F	0.52	SO2P1E	0.56
SAP1H	0.62	SO1P1G	0.71		
				SO2P2A	0.62
SAP2C	0.51	SO1P2E	0.72	SO2P2B	0.65
SAP2E	0.55	SO1P2F	0.52		
SAP2F	0.44	SO1P2G	0.50	SO2P4A	0.53
SAP2G	0.55			SO2P4B	0.47
		SO1P3B	0.63	SO2P4C	0.52
SAP3B	0.51	SO1P3C	0.62	SO2P4D	0.46
SAP3D	0.41			SO2P4E	0.60
SAP3G	0.34	SO1P4A	0.55		
SAP3H	0.33	SO1P4C	0.58	SO2P5A	0.54
SAP3I	0.32	SO1P4D	0.60	SO2P5B	0.30
SAP3J	0.50				
		SO1P5D	0.61		
SAP4A	0.36	SO1P5E	0.42		
SAP4C	0.44				
SAP5B	0.56				
SAP5E	0.43				
SAP5F	0.41				

A.2   Results of footing tests by LEUSSINK *et al.*

Footing on dense sand

Figures A.1-A.3 show load-settlement curves of footing test series A_1 (footing 1×1 m, dense sand $I_n = (n_{max} - n)/(n_{max} - n_{min}) = 0.83 - 0.86$, footing $t = 0.0 - 0.64$ m embedded in soil). The water contents was less than 3%.

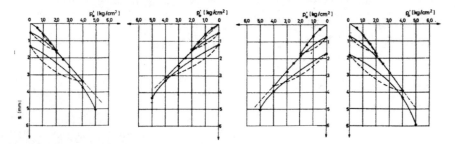

Fig. A.1: Load-settlement curves for test A_1/II, $I_n = 0.83$, $t = 0$ m; dynamometers 1-4 (from left to right), location of dynamometers see Fig. 1.13

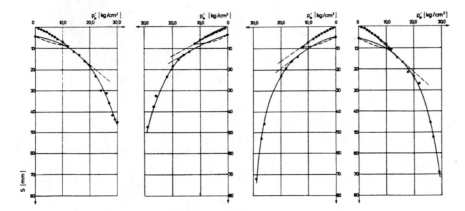

Fig. A.2: Load-settlement curves for test A_1/V, $I_n = 0.86$, $t = 0.25$ m; dynamometers 1-4 (from left to right), location of dynamometers see Fig. 1.13

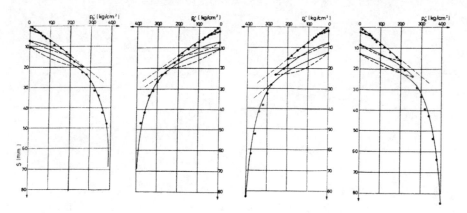

Fig. A.3: Load-settlement curves for test A_1/V, $I_n = 0.86$, $t = 0.64$ m; dynamometers 1-4 (from left to right), location of dynamometers see Fig. 1.13

Footing on very dense sand

Figures A.4-A.6 show load-settlement curves of footing test series A_1 (footing 1×1 m, very dense sand $I_n = 1.01 - 1.18$, footing $t = 0.0 - 0.50$ m embedded in soil). The water contents was less than 3%.

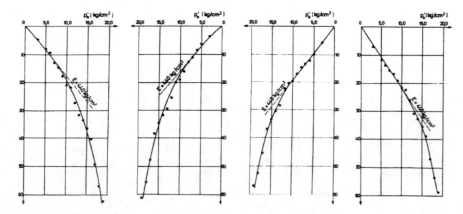

Fig. A.4: Load-settlement curves for test A_1/IX, $I_n = 1.18$, $t = 0.0$ m; dynamometers 1-4 (from left to right), location of dynamometers see Fig. 1.13

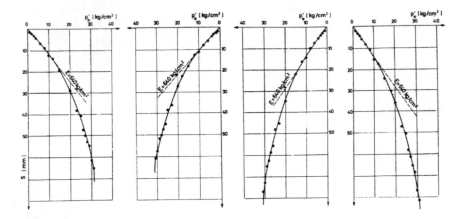

Fig. A.5: Load-settlement curves for test $A_1/VIII$, $I_n = 1.01$, $t = 0.25$ m; dynamometers 1-4 (from left to right), location of dynamometers see Fig. 1.13

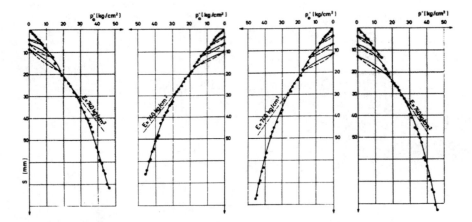

Fig. A.6: Load-settlement curves for test A_1/VII, $I_n = 1.15$, $t = 0.50$ m; dynamometers 1-4 (from left to right), location of dynamometers see Fig. 1.13

A.3 Glossary of symbols

d_{50}	mean grain diameter	C	coefficient of curvature
d_x	grain size, $x\%$ passing sieve		$C = (d_{30})^2/(d_{10} \cdot d_{60})$
d_c	diameter of specimen after	D	diameter of constructive elements
	consolidation	E	YOUNG's (elastic) modulus
e	void ratio	E_s	stiffness modulus
e_0	initial void ratio	F	force, $F = mv^2 = m\omega^2 r$
e_c	void ratio after consolidation	K_0	earth pressure coefficient at rest
e_c	pressure dependent critical void	T	temperature
	ratio	U	coefficient of uniformity
e_{c0}	pressure dependent initial critical		$U = d_{60}/d_{10}$
	void ratio	V	volume
e_d	pressure dependent lowest void	V_0	initial volume
	ratio	V_c	volume of specimen after
e_{d0}	pressure dependent initial lowest		consolidation
	void ratio	ε	strain
e_{max}	max. void ratio	ε_a	axial strain
e_{min}	min. void ratio	$\varepsilon_{a,p}$	axial strain at peak
g	gravitational scaling factor	ε_r	radial strain
h	height	ε_v	volumetric strain
h_c	height of specimen after	$\varepsilon_{v,mp}$	volumetric strain due to mebrane
	consolidation		penetration
h_s	granulate hardness	$\varepsilon_{v,reg}$	registered volumetric strain
m	mass	γ	bulk weight
n	geometric scaling factor	γ_d	dry density
n	factor	$\gamma_{d,c}$	dry density after consolidation
n	exponent, equ. (2.4)	γ_s	specific weight
n	porosity	ν	POISSONS's ratio
n_{min}	min. porosity	ν_{mp}	unit membrane penetration
n_{max}	max. porosity	φ'	friction angle
p'	mean effective stress		$\varphi' = \arcsin(\sigma_1' - \sigma_2')/(\sigma_1' + \sigma_2')_{max}$
	$p' = (\sigma_1' + 2\sigma_2')/3$	φ_{mob}	mobilized friction angle
p'_{eq}	equivalent stress	φ'_p	peak friction angle
q	deviatoric stress	φ'_c	critical friction angle
	$q = \sigma_a - \sigma_r = \sigma_a' - \sigma_r'$	ϱ	specific gravity
r	radius	σ	stress
r_e	pressure dependent relative	$\sigma'_{1,2,3}$	main effective stresses
	density	σ_a'	effective axial stress, for triaxial
t	time		tests $\sigma_a' = \sigma_1'$
t	membrane thickness	σ_r'	effective radial stress, for triaxial
v	velocity, tangential velocity		tests $\sigma_r' = \sigma_2' = \sigma_3'$
	$v = r\omega$	σ_v	vertical stress $sigma_v = \gamma h$
A	area	ψ'_p	dilatancy angle at peak
A_0	initial area	ω	rotational velocity
A_m	surface area of membrane	Π	dimensionless variable